ELECTROSPINNING METHOD USED TO CREATE FUNCTIONAL NANOCOMPOSITES FILMS

Edited by **Tomasz Tański, Pawel Jarka** and **Wiktor Matysiak**

Electrospinning Method Used to Create Functional Nanocomposites Films
http://dx.doi.org/10.5772/intechopen.70984
Edited by Tomasz Tański, Pawel Jarka and Wiktor Matysiak

Contributors

Meikandan Megaraj, Malar Mohan K, Qinglin Wu, Zhen Zhang, Meichun Li, Changtong Mei, Xiuchang Zhang, Tingzhou Lei, Lin Pan, Dinesh Lolla, Harshal Gade, George Chase, Liliana Rozemarie Manea, Petru-Andrei Bertea, Elena Nechita, Carmen Violeta Popescu, Giulia Massaglia, Marzia Quaglio, Okechukwu Ifegwu, László Poppe, Diána Balogh-Weiser, András Szilágyi, Benjámin Gyarmati, Ferenc Ender, Csaba Németh, Tomasz Arkadiusz Tański, Wiktor Matysiak, Pawel Jarka

Notice

First published in London, United Kingdom, 2018 by IntechOpen
IntechOpen is the global imprint of INTECHOPEN LIMITED, registered in England and Wales, registration number: 11086078, The Shard, 25th floor, 32 London Bridge Street
London, SE19SG – United Kingdom
Printed in Croatia

British Library Cataloguing-in-Publication Data
A catalogue record for this book is available from the British Library

Additional hard copies can be obtained from orders@intechopen.com

Electrospinning Method Used to Create Functional Nanocomposites Films, Edited by Tomasz Tański, Pawel Jarka and Wiktor Matysiak
p. cm.
Print ISBN 978-1-78923-580-7
Online ISBN 978-1-78923-581-4

For EU product safety concerns:
IN TECH d.o.o., Prolaz Marije Krucifikse Kozulić 3, 51000 Rijeka, Croatia,
info@intechopen.com or visit our website at intechopen.com.

Meet the editors

Prof. Tomasz Tański is the Head of the Institute of Engineering Materials and Biomaterials, Silesian University of Technology, and a specialist in materials such as nonferrous alloys, in composite and nanostructured materials, in manufacturing engineering, and in surface, properties, and structures of engineering materials. He has authored or co-authored more than 300 scientific publications worldwide including 12 monographs and books and more than 70 publications in the Philadelphia list; won 18 national and international awards and honours; and has served, or is currently serving, as a supervisor or contractor for more than 15 research and didactic projects in Poland and abroad. Prof. Tański is also a reviewer and promoter of numerous scientific papers, including seven doctoral researches in the field of nanotechnology and materials.

Wiktor Matysiak received his PhD degree at the Institute of Engineering Materials and Biomaterials, Silesian University of Technology, where he is currently employed. The subject of his doctoral thesis was "optical and electrical properties of novel types of the polymer nanocomposites with 0D and 1D nanostructures." He specializes in the use of the electrospinning method to produce novel types of thin fibrous mats with special physical properties. During his scientific activity, he was a participant in many scientific international conferences. He is an author and co-author of more than 130 scientific publications worldwide including 17 publications in the Philadelphia list. He is and was a contractor in more than ten research and didactic projects and won 18 national and international awards and honours.

Dr. Paweł Jarka obtained his doctoral degree in Technical Sciences in the field of materials science at the Silesian University of Technology. The subject of his doctoral thesis was "optoelectronic properties of active layers obtained by PVD method for organic photovoltaic cells." He continued to work on organic, polymeric, and nanocomposite-layered structures including the production of micro- and nanofibre mats for applications among others in photovoltaics and optoelectronics. Currently, his scientific interest includes primarily the analysis of optical and electrical properties and the structure of nanostructured materials. He is the author of over 20 scientific articles, including 3 on the Philadelphia list. He also participated in the implementation of numerous research grants.

Contents

Preface

The most effective method of producing nanofibres is the technology of producing in the electrostatic field, which does not require the use of complicated procedures and equipment. Electrospinning allows to produce 1D nanostructures on an industrial scale in a relatively easy and quick way. The method of electrospinning shares the most features with classical technologies in obtaining synthetic fibres that enable forming and generating a stream of previously dissolved or melted polymer and its coaxial stretching, combined with the transition of the polymer from a liquid state to a solid state. In view of the large application possibilities of electrospun fibres, electrospinning is enjoying a dynamically growing interest of scientists, which can be proven by the increasing trend of scientific publications.

This book intends to provide the reader, not only for students but also for professional engineers who are working in the industry as well as for specialists, a comprehensive overview of the state of the art in new trends, research results, and development of electrospinning method of polymer and composite nanofibres and their properties. Chapters for this book have been written by respected and well-known researchers and specialists from different countries. We hope that after studying this book, the reader will have objective knowledge about new aspects in the topic concerning one-dimensional nanostructure electrospinning.

Tomasz Tański, Wiktor Matysiak, and Paweł Jarka
Institute of Engineering Materials and Biomaterials
Silesian University of Technology
Gliwice, Poland

Chapter 1

Introductory Chapter: Electrospinning-smart Nanofiber Mats

Tomasz Tański, Wiktor Matysiak and Paweł Jarka

Additional information is available at the end of the chapter

http://dx.doi.org/10.5772/intechopen.77198

1. Introduction

1.1. Electrospinning: the importance of the method

The available literature describes numerous ways to obtain nanofibers, nanowires or nanorods based on chemical or physical reactions. All of the materials mentioned, that is, nanofibers, nanowires and nanorods, belong to the same one-dimensional group of nanomaterials. Literature reports are not consistent in defining and differentiating these types of materials, yet it can be generally assumed that nanofibers are structures whose length is much larger than the fiber diameter (over 100 times).

They are formed in continuous processes (e.g., during the electrospinning process), as a result of which it is not possible to determine their exact length. Nanowires are considered to be structures of shorter length, in the order of a few nm to several μm.

Shorter than the length of nanowires, nanorods are the structures with the shortest length among the above-mentioned one-dimensional nanomaterials. The main methods of producing one-dimensional nanostructures include template-assisted synthesis, vapor-liquid-solid, physical vapor deposition, magnetron sputtering system, chemical vapor deposition, zol-gel method, molecular self-assembly, nanolithography, and **electrospinning**.

In 2017, over 6000 publications were released worldwide (almost 50% of all publications in the field of nanotechnology published that year), whose topics covered the scope of production methods and analysis of the physical properties (including mechanical, electrical and optical), chemical properties and application possibilities (e.g., in the military, medical or high technology sectors) of polymer nanofibers or composite materials produced with their participation (**Figure 1a**).

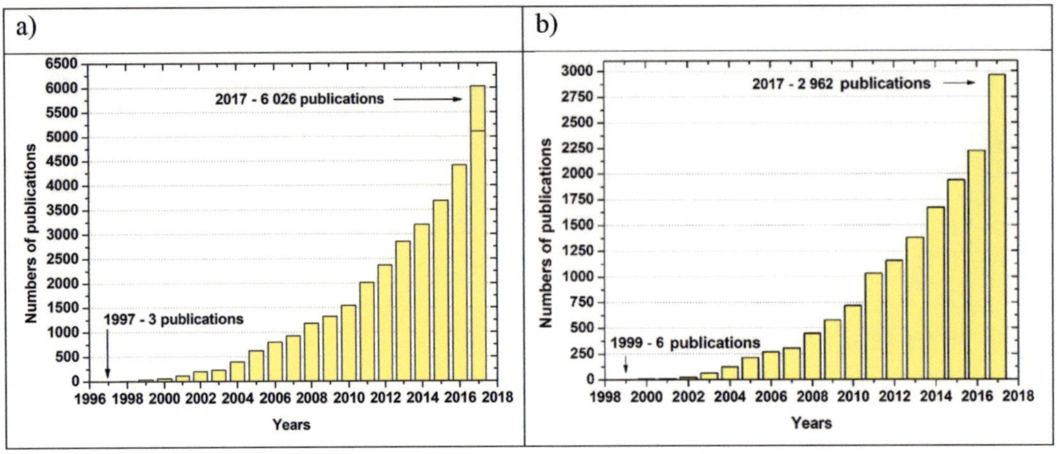

Figure 1. A chart presenting the number of publications in the field of: (a) production and/or testing of nanofibers in 1997–2017; via Scopus and ScienceDirect (keyword: nanofibers, December 2017); (b) electrospinning methods in the years 1998–2017; via ScienceDirect (keyword: electrospinning, December 2017).

A wide range of application possibilities of one-dimensional polymer and composite nanostructures affects the continuous growth of the value of the global market related to products based on nanofibers, which was worth USD 203.2 million in 2013, and a year later it already grew up to USD 276.8 million according to the BCC Research report of May 2016.

It is predicted that this market will increase from USD 383.7 million in 2015 to nearly USD 2 billion in 2020, which will correspond to the annual growth rate (CAGR) at a level of as much as 38.6% between 2015 and 2020.

From the above-mentioned methods, the most effective method of producing polymer and composite nanofibers is the technology of producing fibers in the electrostatic field, which does not require the use of complicated procedures and expensive equipment. This type of process allows to produce one-dimensional polymer and composite nanostructures on an industrial scale in a relatively easy and quick way. Unlike other techniques for the production of nanofibers, the electrospinning method has a significant advantage that, in most cases, the process is carried out at room temperature, and atmospheric pressure. Furthermore, in order to produce nanofibers, only a properly prepared spinning solution, which is usually prepared using a solvent suitable for a given polymer and a simple magnetic stirrer, is required. As a result, it is possible to quickly and cheaply obtain fibrous nanostructures with strictly defined and controlled morphology, desired chemical composition, and structure. An additional advantage of this technology is the fact that in the electrospinning process, it is possible to use most of the polymers known to date in the world, which perfectly illustrates the application possibilities of the technique of electrospinning from the solution.

The method of electrospinning nanofibers shares the most features with classical technologies of obtaining synthetic fibers that enable forming and generating a stream of previously dissolved or melted polymer and its coaxial stretching, combined with the transition of the polymer from a liquid state to a solid state.

Although the technique itself is not new, due to the large application possibilities, it is enjoying a dynamically growing interest of scientists, which can be proved by the increasing trend of scientific publications on the electrospinning method published in the years 1997–2017 (**Figure 1b**).

The method of obtaining polymer fibers using the electrostatic field was patented for the first time by the American professional inventor and electrician, John Francis Cooley, in Great Britain in 1900. In his application, Cooley proposed four technological solutions, which included a standard nozzle, coaxial nozzles, a model in which the polymer bundle was blown through the air stream in addition to interaction with the electrostatic field, and a model with a rotating spinning solution distributor. The next two patent applications, "Apparatus for electrically dispersing fluids" and "Electrical method of dispersing fluids," were registered successively at the US Patent Office in 1902 and 1903. Independently from Cooley, in 1902, an American physicist, William James Morton, registered a patent in which he described a method using a funnel as a kind of spinning nozzle, from which a solution based on nitrocellulose $[C_6H_7(NO_2)_3O_5]_n$ and diethyl ether $C_4H_{10}O$ freely flowed out. The spinning solution flowing out from the funnel was to interact electrostatically with the anode characterized by a spherical shape, and then a cobweb-like mass was to be collected onto the rotating reel. At that time, Morton had already noticed the industrial possibilities of the method of electrospinning from the solution, claiming that "it may be put to any industrial use."

In 1920, another American physicist, John Zeleny, published a work in which he described the behavior of a drop of liquid escaping from the end of a metal capillary under the influence of an electrical voltage. This publication initiated mathematical attempts to analyze and models the behavior of a stream of liquid in an electrostatic field. So far, numerous mathematical models have been developed describing the behavior of the spinning solution bundle in the electrostatic field during the electrospinning process. However, none of the models described is accurate enough so that the theoretical results obtained overlap with experimental data.

In the period from 1964 to 1969, a British physicist, Sir Geoffrey Ingram Taylor, carried out scientific work on the conduct of conductive liquids in the electrostatic field. His work describing the mathematical modeling of the change in the shape of a spinning solution drop coming out of the nozzle due to the applied potential difference significantly contributed to the development of the method of obtaining polymer nanofibers by means of the technique of electrospinning from the solution. A characteristic conical shape that a spinning solution drop takes during the electrospinning process is now known as the so-called Taylor cone. In 1971, Peter K. Baumgarten conducted research on the production of polymer nanofibers by means of the method of electrospinning from the solution based on acrylic resin and dimethylformamide. The analysis of high-speed photography (images taken with very short exposure time) of a stream of spinning solution in the electrostatic field showed that although the electrospinning process is observed as a "hazy cloud" resembling a spinning solution bundle on the way between the nozzle and the collector, only one fiber is formed during the electrospinning process. Through interaction with the electrostatic field produced between the electrodes, this fiber moves in a spiral motion and then settles on the surface of the grounded collector in the form of a fibrous mat. In addition, calculations based on the photographs of the polymer bundle during the electrospinning process showed that the resulting fiber on the section between the nozzle and the collector probably moves at speed exceeding the speed of sound in the air.

Scientific works in the field of the process of electrospinning polymer nanofibers in the electrostatic field conducted in the 1990s of the twentieth century by research groups under the direction of an American physicist, Dr. Darrell H. Reneker, have shown that it is possible to produce fibers from numerous organic polymers. This fact has become the reason for the dynamic growth of interest in the electrospinning method observed to this day.

During the electrospinning process, an electrostatic field is generated under the influence of a high voltage of several to several dozens of kilovolts between the nozzle (a metal needle of a syringe to which the spinning solution is delivered at a constant speed) and a grounded collector (**Figure 2**).

The presence of a potential difference between the electrodes due to electrostatic interactions causes the electric charges to be induced on the surface of the spinning solution drop emerging from the nozzle. A negatively charged collector located under the nozzle causes the repulsion of negative charges of the solution drop toward the soul, which additionally attracts them with Coulomb forces due to its positive resultant charge, as a result of which positive charges are accumulated on the surface of the spinning solution drop emerging from the nozzle. Under the influence of electrostatic field forces caused by a correspondingly large potential difference between the electrodes, the drop of solution at the mouth of the nozzle opening becomes distorted and adopts a conical shape (the so-called Taylor cone), which is accompanied by the movement of charges carried by a stream of the spinning solution toward the grounded collector. The outflow of spinning fluid toward the collector is initiated by exceeding the critical intensity of the electrostatic field, and then a drop of the polymer solution is stretched to a thin fiber under the influence of field forces. When the diameter of

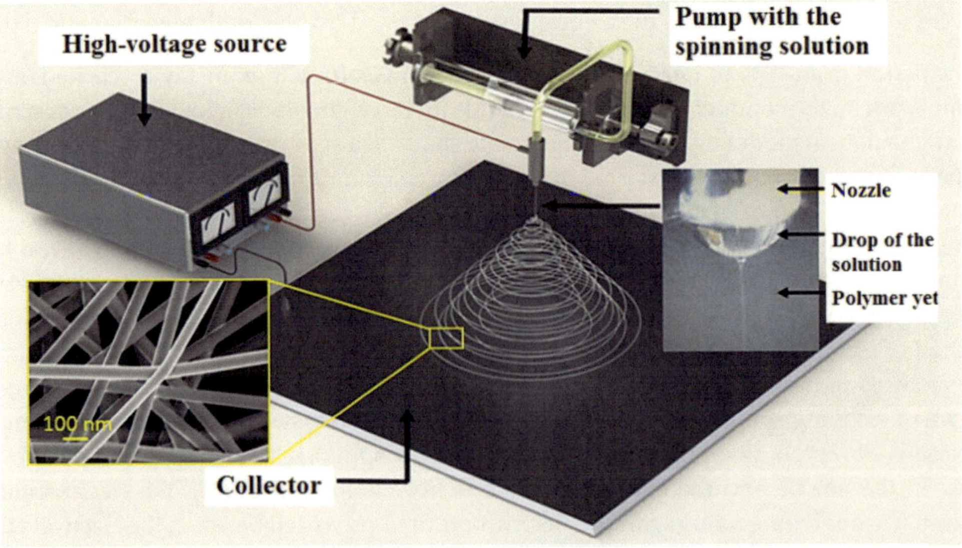

Figure 2. A diagram of a stand for the production of nanofibers containing a picture showing the mouth of the nozzle and a drop of the polymer solution that forms under the influence of high voltage in the so-called Taylor cone from the apex, which is followed by a shot of a thin stream of the solution.

the stream decreases, the ratio of the stream surface to its volume increases. The solvent evaporates, and the spinning solution solidifies to a form of a polymer fiber, which then settles on the surface of the conductive collector.

The morphology and properties of polymer nanofibers obtained by means of the electrospinning method are influenced by many factors that can be divided into three main groups: spinning solution parameters, apparatus parameters, and environmental parameters.

The main parameters resulting from the type of spinning solution used include: solution viscosity, polymer mass concentration relative to the solvent used, the molecular weight of the polymer used, type of solvent, electrical conductivity, and surface tension of the produced solution. In addition, the key parameters that have a significant impact on both the morphology and physical properties of the fibers produced are process parameters used during electrospinning. These parameters include the speed of feeding the spinning solution to the mouth of the nozzle, the difference in potentials, and the distance between the electrodes, that is, the nozzle and the collector. These process parameters can be changed during the electrospinning process, which makes it possible to control their values in order to obtain a stable Taylor cone formed from a spinning solution drop at the mouth of the nozzle. In addition, the process parameters also include apparatus parameters, taking into account the inner diameter, and the length of the nozzle used as well as the type of collector. We can distinguish two basic types of collectors used during the electrospinning process of nanofibers. They include flat plate collectors and drum collectors, which rotate at a precisely defined angular speed during the electrospinning process, thus winding the produced fibers as on a reel. As a result, nanofibers with a chaotic arrangement are obtained using a flat plate collector, while the use of a drum collector allows for obtaining nanofibers with a strictly defined orientation and a parallel arrangement of individual fibers relative to each other. The environmental parameters include ambient conditions in which nanofibers are produced, that is, temperature, air humidity, and atmospheric pressure.

Author details

Tomasz Tański*, Wiktor Matysiak and Paweł Jarka

*Address all correspondence to: tomasz.tanski@polsl.pl

Institute of Engineering Materials and Biomaterials, Silesian University of Technology, Gliwice, Poland

Chapter 2

Fabrication of a Superhydrophobic Nanofibers by Electrospinning

Meikandan Megaraj and Malarmohan Keppannan

Additional information is available at the end of the chapter

http://dx.doi.org/10.5772/intechopen.75357

Abstract

The major work of this research work is fabrication and investigation of the surface characteristics of a hydrophilic polycaprolactone (PCL) and superhydrophobic 1H, 1H, 2H, 2H-perfluorodecyltriethoxysilane (PFDTES)-$AgNO_3$-modified PCL fibrous membranes through the electro spinning technique. The surface properties of the PCL fibrous were modified from hydrophilic to superhydrophobic by adding (0.05 vol%) PFDTES-$AgNO_3$ solution into a mixed with a solvent of PCL and chloroform. The electrospun PFDTES-modified PCL fibrous showed a maximum water contact angle (WCA) of 158° due to increase in surface roughness when compared with the PCL fibrous roughness, having a maximum WCA of 81° and an average fiber diameter of 400–700 nm.

Keywords: electrospinning, polycaprolactone (PCL), nanofiber, water contact angle (WCA), superhydrophobic

1. Introduction

The superhydrophobic surface normally states, mixture of the static contact angle more than 150° with a contact angle hysteresis lesser than 5°. It is categorized into two types agreeing to water rolling angle; an enormously adhesive superhydrophobic surface that permits water droplets to adhere the surface, even when the surface is turned upside down and less adhesive superhydrophobic surface with a rolling angle less than 10° [1–3]. Electrospinning has been keenly exploited as a simple and flexible method for producing ultrathin fibers made of several materials. Countless advancement had been made in latest years with concern to the theory for electrospinning and mechanism of the orientation of electrospun fibers [4]. Electrospinning is an efficient and simple method for fabrication of constant nanofibrous with high surface bumpiness and the surface-to-volume ratio of the driving force of an outward electric field on polymer solutions or polymer liquefies [5]. Since polymeric structures allow the numerous everyday applications areas of nanoparticles are as additives to polymers used in the motorized and aerospace division were vehicle parts for lesser weight and greater performance, packaging including food and biomedical to keep and preserve the reliability of the product by controlling the obstacle, mechanical, optical and respiration properties, in textiles industries increases in strength, water resistance, self-cleaning, fade resistance and special care merchandises UV shield, deep dispersion skin cream emulsions. Many of these functionalities can be exchanged from one use to another. For example, the similar technology used for transparent UV protective coatings such as sunscreens in personal care products can be used for UV protection in food packaging, paints, and textiles [6]. Coming together or altering the basic frame with another component for the electrospinning applications, one can produce modified nanocomposite hybrid fibers for the variety of day to day applications. So far, a number of techniques have been effectively expressed for generating rough surface structures. Among others, electrospinning as a low-cost, continual, scalable nanomanufacturing technique has been widely engaged for fabricating continuous nanofibers/microfibers of a huge variety of natural and synthetic polymers, polymer derived carbon, metals, metal oxides and ceramics, etc. [7–12].

There have also been many noteworthy models describing the production of low energy hydrophobic and superhydrophobic surfaces by electrospinning. Jiang et al. produced superhydrophobicity of Fe_3O_4-filled carbon nanofibers through sintering electrospun [13]. Cao et al. fabricated a superhydrophobic surface on calcined electrospun SiO_2 nanofibers using a tridecafluoro-1,1,2,2- tetrahydrooctyldimethyl-chlorosilane layer [14]. Ma et al. defines low surface energy fibers produced by electrospinning polystyrene—polydimethylsiloxane copolymers which were unclean with polystyrene homopolymer. Relatively more content of low surface energy siloxane polymer caused in fibers with water contact angles of 163° [15]. Rutledge et al. stated that chemical vapor deposition coating of polytetrafluoroethylene having superhydrophobic properties [16]. Miyauchi et al. fabricated a biomimetic superhydrophobic surface including micro-nanoporous poly styrene microfibers by the use of electrospinning technique [17]. Jiang et al. electrospun polystyrene from DMF/THF to give a fiber mat with contact angles of 139.1° and went on to spin/spray a dilute solution of polystyrene to give a film of porous microparticles which gave a water contact angle of 162° [18]. Acatay et al. used a fluorinated comonomer, present at up to 50 wt% to attain superhydrophobic surfaces via an electrospinning process using a copolymer of acrylonitrile and α,α-dimethyl meta-isopropenyl benzyl isocyanate and 50 wt% of a perfluorinated diol [19]. Borner et al. used a one-step process to produce poly(lactic-co-glycolic acid) nanofiber interconnects with surfaces improved biofunctional peptides by spinning a homogeneous mixture of PLGA and a polymer-peptide conjugate [20]. Bianco et al. spun polyamide 6 nanofibers in the occurrence of fluorinated acridine. They identified that the addition of increasing volumes of the acridine (2–6 wt%), static contact angles with water on the fibers improved gradually from 62° for unchanged polyamide to 123° [21]. Considering the fascinating features of superhydrophobic surface an attempt is made in the present work to develop a superhydrophobic surface of polycaprolactone and PFDTES-modified polycaprolactone nanofibrous through an electrospinning technique. The surface characterization studies such as morphological feature, chemical composition, and water contact angle were analyzed and reported.

2. Experimental details

The reagents used in the present study were polycaprolactone (PCL), chloroform, 1H, 1H, 2H, 2H-perfluorodecyltriethoxysilane (PFDTES), ethanol and acetone. All other chemicals were of analytical grade and were used as received purchased from Sigma Aldrich, India.

Electrospinning is a simple and flexible method for producing ultrathin fibers. Altering the basic frame with another component for electrospinning, can produce modified nanocomposite hybrid fibers for various applications. So far, a number of techniques have been effectively expressed for generating rough surface structures. Among others, electrospinning is a low-cost, continual, scalable nanomanufacturing technique and is widely used for fabricating continuous nanofibers/microfibers of a huge variety of natural and synthetic polymers; polymer derived carbon, metals, metal oxides and ceramics, etc.

The reagents used in the present study are polycaprolactone (PCL), silver nitrate ($AgNO_3$), chloroform, ethanol, PFDTES, and acetone. All other chemicals are of analytical grade and purchased from Sigma Aldrich, India. The photographic and schematic view of electrospinning setup that was used in this preparation method is shown in **Figure 1(a)** and **(b)**, it consists

(a)

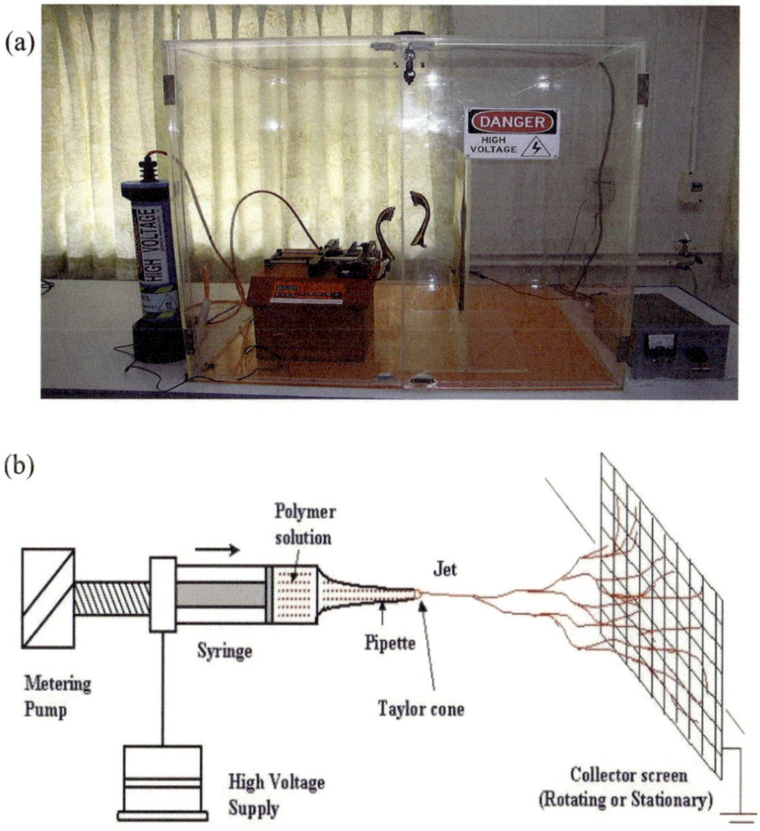

Figure 1. (a) Photographic view of electrospinning setup and (b) schematic view of electrospinning setup.

of a high-voltage supplier, grounded target board and a syringe pump. The polymer solution was flowing through a needle of 1 mm inside diameter. Solutions were prepared by liquefying the desired amount of polymer (according to concentration) in a solvent (chloroform). Solutions are prepared in beakers. Glass beaker was protected with aluminum foil and plastic sheet throughout the process of suspension to prevent solvent evaporation. The process of liquefying was continued using magnetic stirrer. PFDTES-$AgNO_3$-modified PCL solutions are prepared in the same way which was followed by the addition of PFDTES.

Initial experiments were carried out to find best conditions for electrospinning PCL and PFDTES. $AgNO_3$ modified PCL fiber membranes were obtained by varying solution concentration (5–15% W/V), feed rate (5, 10, 15, and 20 ml/h) and voltage (5, 15, and 25 kV). The needle to tip collector distance (NCD) was 10 cm and a 10 ml syringe fitted with a 0.838 mm inside diameter and stainless steel needle (Sigma Aldrich) was used. PCL and (0.05% W/V) PFDTES-$AgNO_3$-modified PCL nanofibers were prepared with electrospinning technique with 10 & 15% W/V concentration using chloroform as an organic solvent. The applied dc voltage was kept at 10 kV, tip target distance was maintained at 10 cm and the flow rate was kept at 0.01 ml/min.

3. Results and discussion

3.1. Nanofibers by electrospinning

The solvent used for electrospinning plays a vital role in the morphology of the subsequent electrospun polymer fibers. Nanofibers achieved by electrospinning frequently exhibit beaded fiber structures, which are significantly influenced by the solution properties. Initially, the concentration of polymer solution plays a significant role in the formation of beads. At dilute concentration, mostly beads are produced because of lack of sequence entanglement in polymer solution, beaded fiber structure is generated in the medium concentration. Normally, beaded fibers have been considered as unwanted or faulty products. Therefore, at a concentrated solution the continuous fiber structure, without bead is attained. The morphology of electrospun PCl fiber membrane and PFDTES-$AgNO_3$-modified PCL fiber membrane is shown in **Figures 2–4**. The PCL fibers and PFDTES-$AgNO_3$-modified PCL fibrous membrane, having widely distributed fiber diameters, were randomly oriented as a porous membrane. Furthermore, the fibers exhibited smooth fibers and they have an average fiber diameter of 400–700 nm.

The wettability of prepared substrate was measured, in relations of contact angle, by means of a Goniometer based on sessile drop technique. A drop size of 5 μl was used and the contact angle measurement was done at six different locations in a sample conserved at room temperature. A water droplet placed on PCL fibrous membrane is shown in **Figure 5**. The water droplet was straightaway absorbed by this fibrous membrane.

After PFDTES-$AgNO_3$ modification, the PCL fibers membrane still maintained the fiber shape. Smooth fibers observed from **Figure 6(a)** and **(b)** show a water droplet placed on PFDTES-$AgNO_3$-modified PCl fiber membranes. It can be found that a high surface hydrophobicity WCA lies in between 150 and 158° of fibrous membranes was obtained after the PFDTES-$AgNO_3$ modification.

Figure 2. Influence of PCL solution concentration of 15% W/V observed under SEM.

Figure 3. Influence of PFDTES-$AgNO_3$-modified PCL solution concentration of 15% W/V observed under SEM.

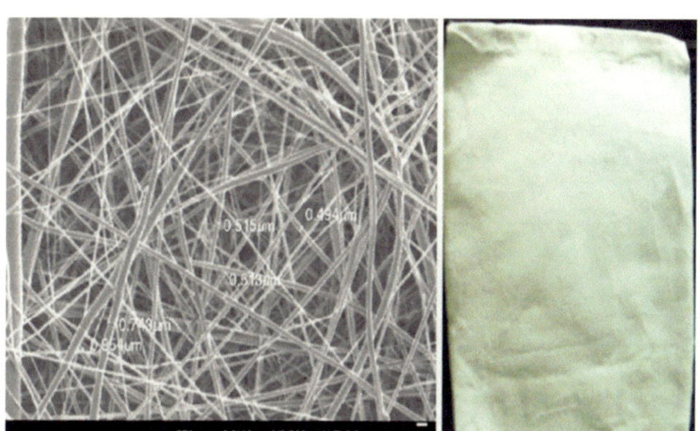

Figure 4. Influence of PFDTES-$AgNO_3$-modified PCL solution concentration of 10% W/V observed under SEM.

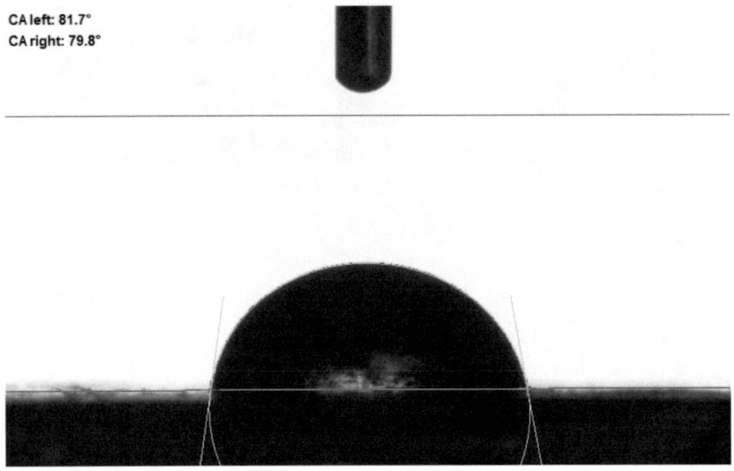

Figure 5. Water droplet on PCL fibers membrane.

The PFDTES-$AgNO_3$ modification it was proved to be effective to increase the membrane hydrophobicity. PFDTES is a well-known low-surface-energy material. From the above results, the surface coated with PFDTES improves the superhydrophobicity through the formation of micro/nano metric arrangement and these structures are stable. However, entrapment of air between the fibrous membranes allows the water drops to roll off the surface with ease.

Surface profilometer top view images (10 × 10 μm) and cross-section profiles of PFDTES-$AgNO_3$-modified PCL fibers membrane are displayed in **Figures 7–9**. The electrospun fibrous membrane shows high surface roughness due to the random deposition of the fibers on the collector end.

The stability of hydrophobicity nature aimed at PFDTES-$AgNO_3$-modified PCL fiber membranes were deliberate and analyzed under ambient, low 10°C and high temperature 110°C

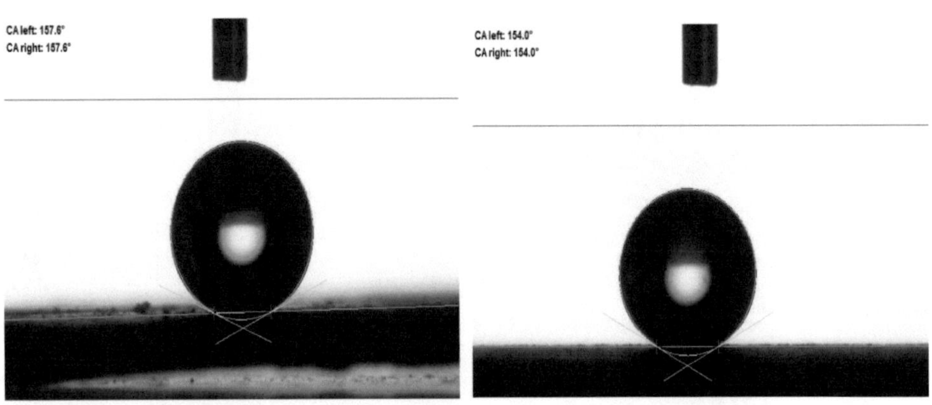

Figure 6. (a and b) Water droplet placed on PFDTES-$AgNO_3$-modified PCL membrane.

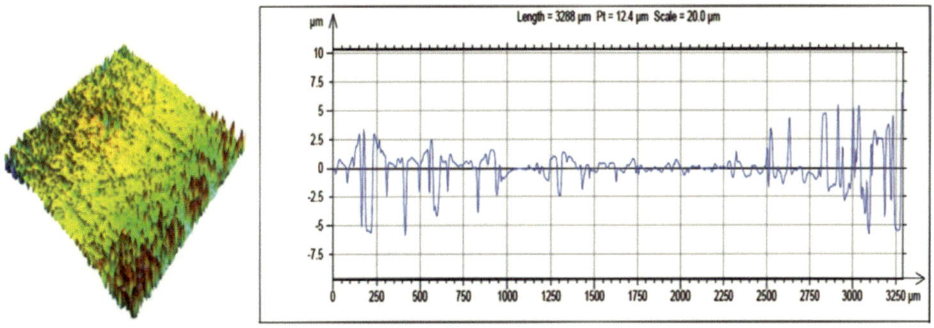

Figure 7. PFDTES-$AgNO_3$-modified PCL fibers membrane with roughness value of 1.52 μm.

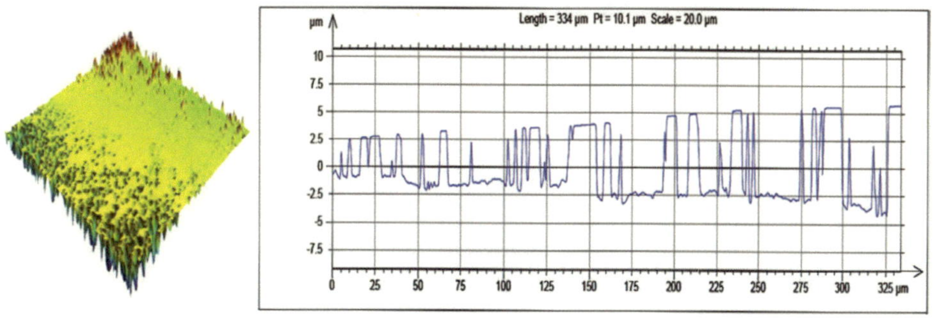

Figure 8. PFDTES-$AgNO_3$-modified PCL fibers membrane with roughness value of 1.85 μm.

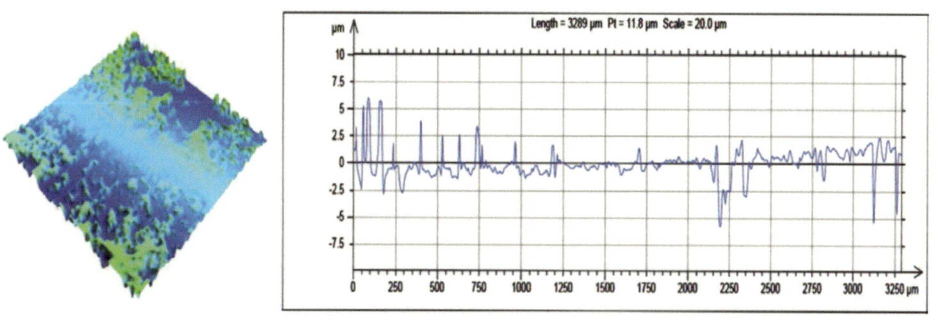

Figure 9. PCL fibers roughness value is 0.82 nm.

conditions for 60 days correspondingly. It is resolved from the **Figure 10** that for all the given temperatures, hydrophobicity nature for PFDTES-$AgNO_3$-modified PCL fiber membrane reduced only by 7°. Based on the discussion, it is understood that the proposed electrospinning is best for developing hydrophobic/superhydrophobic surface. However, after the stability tests at various temperatures it is observed flaking or pops on the top layer of nanofibers. Unfortunately the fibers flake out with the simplest touch or with simple air pressure, and results in dust over the system.

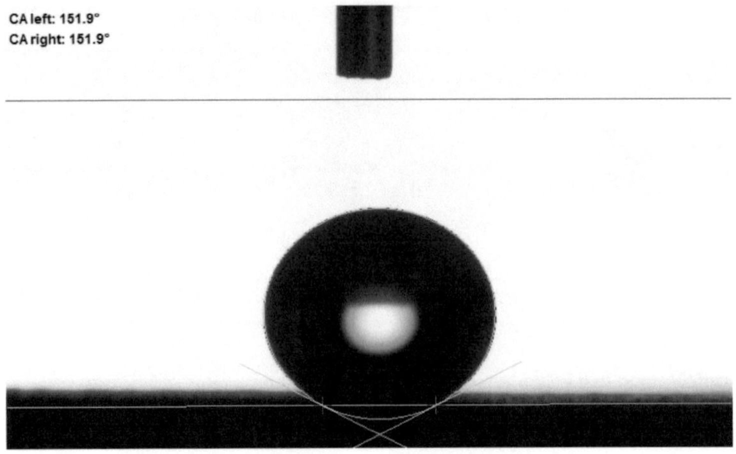

Figure 10. Superhydrophobic nature of PFDTES-$AgNO_3$-modified PCL fibers after durability test.

4. Conclusion

The hydrophilic and superhydrophobic fibrous membranes were fabricated by electrospinning using polycaprolactone (PCL) and 1H, 1H, 2H, 2H-perfluorodecyltriethoxysilane (PFDTES)-modified polycaprolactone. Surface roughness of the PFDTES-modified polycaprolactone nanofibers was controlled using a variety of chloroform solvent mixtures. The addition of PFDTES to the solvent mixture affected the surface morphology of polycaprolactone fibrous and enhanced the WCA of the polycaprolactone fibrous without altering the smoothness of fibers. The electrospun PFDTES-modified polycaprolactone fibrous showed a maximum water contact angle (WCA) of 154° due to increase in surface roughness when compare with the polycaprolactone fibrous roughness, which showed a maximum WCA of 81° and an average fiber diameter of 400–700 nm.

4.1. Drawbacks

- Although electrospinning coating technique achieved enhancement in heat transfer, their durability still needs to be improved to prevent observed flaking or pops after the heat transfer tests.
- Polymer coatings with higher stability on substrates have larger thickness. It involves in larger thermal resistance of the substrate, hence heat transfer characterizes values were not reported.

Author details

Meikandan Megaraj[1,2]* and Malarmohan Keppannan[1,2]

*Address all correspondence to: meikandan013@gmail.com

1 Vel Tech Rangarajan Dr. Sagunthala R&D Institute of Science and Technology, Chennai, India

2 Department of Mechanical Engineering, Anna University, Chennai, India

References

[1] Wu R, Chao G, Jiang H, Hu Y, Pan A. The superhydrophobic aluminum surface prepared by different methods. Materials Letters. 2015;**142**:176-179

[2] Hardman SJ, Muhamad-Sarih N, Riggs HJ, Thompson RL, Rigby J, Bergius WNA, Hutchings LR. Electrospinning superhydrophobic fibers using surface segregating end-functionalized polymer additives. Macromolecules. 2011;**44**:6461-6470. DOI: 10.1021/ma200852z

[3] Meikandan M, Malarmohan K, Velraj R. Development of superhydrophobic surface through facile dip coating method. Digest Journal of Nanomaterials and Biostructures. 2016;**11**:945-951

[4] Diaa BM, Jaafar HT. Superhydrophobic nanocomposites coating using electrospinning technique on different materials. International Journal of Applied Engineering Research. 2017;**12**(24):16032-16038. ISSN 0973-4562

[5] Ding Y, Hou H, Zhao Y, Zhu Z, Fong H. Electrospun polyimide nanofibers and their applications. Progress in Polymer Science. 2016;**61**:67-103. DOI: 10.1016/j.progpolymsci.2016.06.006

[6] Ding B, Li C, Hotta Y, Kim J, Kuwaki O, Shiratori S. Conversion of an electrospun nanofibrous cellulose acetate mat from a super-hydrophilic to super-hydrophobic surface. Nanotechnology. 2006;**17**:4332-4339. DOI: 10.1088/0957-4484/17/17/009

[7] Ogawa T, Ding B, Sone Y, Shiratori S. Super-hydrophobic surfaces of layer-by-layer structured film-coated electrospun nanofibrous membranes. Nanotechnology. 2007;**18**:165607. DOI: 10.1088/0957-4484/18/16/165607

[8] Guo Z, Yang F. Surfaces and Interfaces of Biomimetic Superhydrophobic Materials, Chapter 5. New Jersey, USA: John Wiley & Sons; 2017

[9] Ahmed B, Mohamed AMA, Aboubakr A, Mariam AM'a. Corrosion protection of electrospun PVDF-ZnO superhydrophobic coating. Surface and Coating Technology. 2015;**289**: 946-957. DOI: 10.1016/j.surfcoat.2015.12.087

[10] Huang ZM, Zhang YZ, Kotaki M, Ramakrishna S. A review on polymer nanofibers by electrospinning and their applications in nanocomposites. Composites Science and Technology. 2003;**63**:2223-2253. DOI: 10.1016/S0266-3538(03)00178-7

[11] Dzenis Y. Material science. Spinning continuous fibers for nanotechnology. Science. 2004; **304**:1917-1919. DOI: 10.1126/science.1099074

[12] Jin S, Park Y, Park CH. Preparation of breathable and superhydrophobic polyurethane-electrospun webs with silica nanoparticles. Textile Research Journal. 2016;**86**(17):124-129

[13] Jiang L, Zhao Y, Zhai J. A lotus-leaf-like superhydrophobic surface: A porous microsphere/nanofiber composite film prepared by electrohydrodynamics. Angewandte Chemie International Edition. 2004;**43**:4338-4341. DOI: 10.1002/anie.200460333

[14] Shang HM, Wang Y, Takahashi K, Cao GZ, Li D, Xia YN. Nanostructured superhydrophobic surfaces. Journal of Materials Science. 2005;**40**:3587-3591. DOI: 10.1007/s10853-005-2892-9

[15] Ma M, Hill RM, Lowery JL, Fridrikh SV, Rutledge GC. Electrospun poly (styrene-block-dimethylsiloxane) block copolymer fibers exhibiting superhydrophobicity. Langmuir. 2005;**21**:5549-5554. DOI: 10.1021/la047064y

[16] Gleason KK, Rutledge GC, Gupta M, Ma M, Mao Y. Superhydrophobic fibers produced by electrospinning and chemical vapor deposition. U.S. Pat. Appl. Publ. 2007. p. 26

[17] Miyauchi Y, Ding B, Shiratori S. Fabrication of a silver-ragwort-leaf-like super-hydrophobic micro/nanoporous fibrous mat surface by electrospinning. Nanotechnology. 2006;**17**:5151-5156. DOI: 10.1088/0957-4484/17/20/019

[18] Zhu Y, Jing CZ, Zhai J, Yong MZ, Feng L, Jiang L. Multifunctional carbon nanofibers with conductive, magnetic and superhydrophobic properties. Chemphyschem. 2006;**7**:336-341. DOI: 10.1002/cphc.200500407

[19] Park Y, Park CH, Kim J. A quantitative analysis on the surface roughness and the level of hydrophobicity for superhydrophobic ZnO nanorods grown textiles. Textile Research Journal. 2014;**84**:1776-1788

[20] Gentsch R, Pippig F, Schmidt S, Cernoch P, Polleux J, Borner HG. Single-step electrospinning to bioactive polymer nanofibers. Macromolecules. 2011;**44**:453-461. DOI: 10.1021/ma102847a

[21] Bianco A, Iardino G, Bertarelli C, Miozzo L, Papagni A, Zerbi G. Modification of surface properties of electrospun polyamide nanofibers by means of a perfluorinated acridine. Applied Surface Science. 2007;**253**:8360-8364. DOI: 10.1016/j.apsusc.2007.04.003

Chapter 3

The Place of Electrospinning in Separation Science and Biomedical Engineering

Okechukwu Clinton Ifegwu and Chimezie Anyakora

Additional information is available at the end of the chapter

http://dx.doi.org/10.5772/intechopen.77221

Abstract

Electrospunnanofibers have found myriad of applications from separation science to clinical translation. Electrospunnanofiber scaffolds have the benefits of unique properties such as high surface area to volume ratio, interfibrous pore sizes, strong penetrability, great deal of active sites for adsorption, excellent stability, better targeting, minimum toxicity, high drug-loading capacity, exceptional mechanical properties, flexibility in surface functionality, ease of encapsulation of drugs and bioactive compounds, suitability for thermos-liable drugs, enhanced cellular interactions, and protein absorption to facilitate binding sites for cell receptors. In the field of separation science, electrospunnanofiber scaffolds have extensively served as sorbent material for solid phase extraction techniques mainly due to the need to improve sorptive capacity and analyte selectivity. Given that almost all of the human tissues and organs are deposited in nanofibrous forms or structures, electrospunnanofibers/nanocomposites are currently being investigated for potential clinical applications. It is noteworthy that the nanofiber fabrication technique and the material integrity are key components to obtaining clinically relevant nanofibers. Owing to the significance of fiber arrangement to nanofiber performance, electrospinning has a leading edge over other nanofiber fabrication techniques due to the ease of controlling fiber orientation, despite the inherent advantages of other conventional nanofiber fabrication techniques. The current review highlights the superb qualities of electrospunnanofibers, their various methods of fabrication, and their various applications especially in separation science and clinically. We further provided an overview of the electrospinning principles, types of electrospinning, parameters that affect the nanofibers fabrication via electrospinning, challenges, and the future directions. The advent of robotics-assisted electrospinning technique offers new opportunities for the traditional biofabrication in higher accuracy and controllability and hence will certainly drive nanotechnology from laboratory/industry toward patient care in the near future.

Keywords: electrospinning, separation science, biomedicine, nanoscience

1. Introduction

The emergence of nanotechnology and the unique properties of nanoscale materials have rekindled a renewed interest in the scientific community. The advent of nanotechnology has birthed various functional materials with highly ordered hierarchical structures and superb attributes that can successfully mimic the sophisticated biological processes. An important and exciting direction of research in nanomedicine would be to gain an in-depth understanding and exploit the cellular response to nanostructures. For example, electrospunnanofibers have diameter in the nanometer range, are arbitrarily long, and possess larger surface area (to volume ratio), interfibrous pore sizes, strong penetrability, great deal of active sites for adsorption, and better interaction with other compounds as compared to their microfibers counterpart [1, 2]. They have excellent stability, better targeting, minimum toxicity, high drug-loading capacity, exceptional mechanical properties, flexibility in surface functionality, and encapsulation of drugs and bioactive compounds and are suitable for thermos-liable drugs [1–4]. Endowed with both topographical and biochemical signals such electrospun nanofibrous scaffolds may provide an optimal microenvironment for the seeded cells. As such, electrospunnanofiber scaffolds have found promising applications in various clinical fields. Some notable applications include tissue engineering, regenerative engineering, separation science, biosensors, filtration, wound dressings, drug delivery, and enzyme immobilization, among others [5–7]. Almost all of the human tissues and organs are deposited in nanofibrous forms or structures, such as skin, bone, dentin, collagen, and cartilage to mention but a few. Until recently (5–10 years), electrospinning did not elicit widespread interest as a potential polymer-processing technique for applications in tissue engineering and drug delivery despite the 30 years long history of tissue engineering [8, 9]. This renewed interest was instigated by the aforementioned properties that afford the opportunity to engineer scaffolds with micro- to nanoscale topography and high porosity similar to the natural extracellular matrix (ECM). Electrospinning also makes it possible to incorporate the benefits of nanoparticles in nanofibrous form thus addressing some or all of their limitations [9, 10]. The fibrous and continuous nature of electrospunnanofibers as opposed to nanoparticle expands the possibilities for use in separation science and analytical and clinical applications. The generated fibers have a high surface area to volume ratio; the fibrous mats are highly porous and display excellent mechanical properties when compared to other materials of the same scale. Interestingly, the characteristic high surface area to volume ratio of electrospunnanofiber scaffolds does enhance cell attachment, drug loading, and mass transfer properties [1, 2, 8–10]. Consequently, business opportunities for nanostructured materials in biomedical applications were estimated as on 2006 by Ramakrishna et al. and Teo et al. to be of the order of 180 billion US dollars in 2015 [11, 12].

1.1. Fabrication of electrospunnanofibers

The synthesized nanofibers can be tailored for specific purposes. Several techniques based on different physical principles have been reported in the fabrication of nanofibrous scaffolds with unique properties [13, 14]. The most common ones include drawing [15], template synthesis [16], self-assembly [17, 18], phase separation [13, 18], melt blowing [19–22], and electrospinning [12]. More so, in an attempt to construct nanostructured materials with remarkable biomimetic properties, four bioinspired strategies have emerged from the natural biomineralization process: biostructure mimicking, biofunction anchoring, biotemplating, and bioassembling [23].

1.1.1. *Drawing*

The tip of an atomic force microscope (AFM) or a micropipette is dipped into a droplet of a viscoelastic solution near the contact line using a micromanipulator. Pulled nanofibers are formed by slowly withdrawing the micropipette or AFM tip at a speed of approximately 1×10^{-4} ms^{-1} and are deposited on the surface by touching it with the end of the micropipette. The micropipette or the AFM tip is smoothly withdrawn slowly from the solution (**Figure 1**). The viscosity of the material at the edge of the droplet is usually increased with evaporation. Hence, drawing a fiber requires a viscoelastic material that can undergo strong deformations while being cohesive enough to support the stresses developed during pulling. If the viscoelastic solution is too light, the drawn fiber will break due to Rayleigh instability. Alternatively, if the solution is concentrated at the edge of the droplet, the drawn fiber will break in a cohesive manner.

The drawing method had been successfully employed in fabricating sodium citrate nanofibers using chloroauric acid as the solvent [16]. With this method, it is possible to fabricate fibers with fiber diameters between 2 and 100 nm and fiber length of 10 microns to mms. The drawing process gives a good level of repeatability, convenient to process with minimum equipment requirement but control of fiber dimensions is limited. It is a discontinuous process that cannot be scaled up, and thus it is only suitable for laboratory production of nanofibers. The drawing process can be considered as dry spinning at a molecular level.

1.1.2. *Template synthesis*

Template synthesis implies the use of a template or mold to obtain a desired material or structure (see **Figure 2**). The template refers to a metal oxide membrane with through thickness pores of nanoscale diameter and for the creation of nanofibers with specified diameter the polymer solution is extruded through a template of an appropriate diameter [17]. The

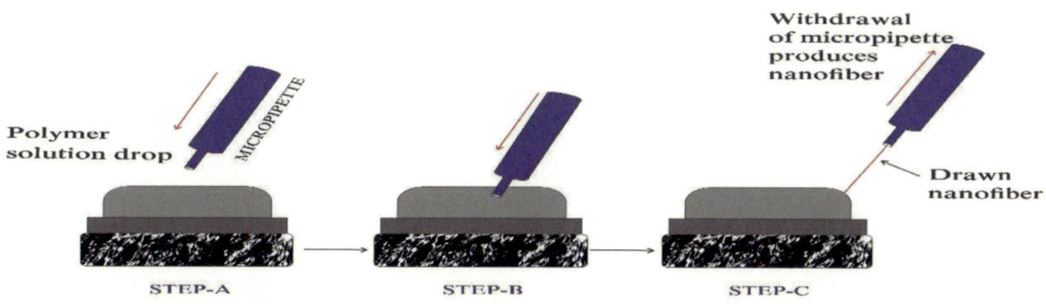

Figure 1. Drawing process in which the micropipette that was in contact with a polymer droplet (step B) is drawn away to generate nanofibers (step C).

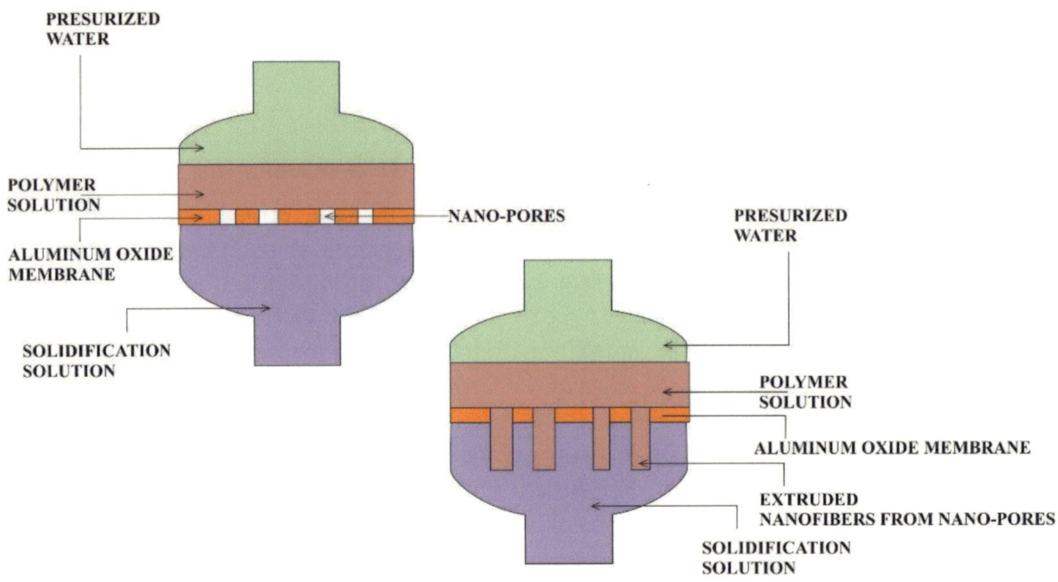

Figure 2. A typical process of obtaining nanofibers by template synthesis.

principle requires that water pressure be applied on one end while the porous end is restrained as this results in extrusion of the polymer melt/solution. Nanofibers are formed once the polymer extrusion is exposed to the solidifying solution and their diameters are determined by the pore sizes [17]. Feng and his coworkers successfully fabricated polyacrylonitrile (PAN) nanofibers of 100 nm fiber diameter and 10 micron fiber length in dimethylformamide (DMF) solvent employing this method [17]. A major advantage of this method is that fibers of different diameters can be easily achieved by using different templates.

1.1.3. Phase separation

In phase separation, a polymer solution is allowed to form a gel and then the solvent is extracted leaving behind the residual porous solid phase (**Figure 3**). The main mechanism

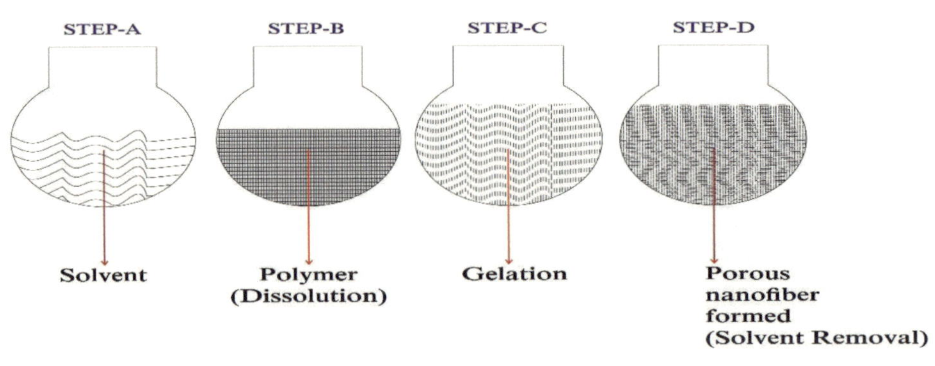

Figure 3. A phase separation process in which the successful gelation (step C) of a polymer solution (step A) and extracting the solvent result in the formation of nanofibers (step D).

in this process is-as the name suggests-the separation of phases due to physical incompatibility. One of the phase-which is that of the solvent-is then extracted, leaving behind the other remaining phase. In 1999, Ma and Zhang [24] had successfully produced 50–500 nm size nanofibrous poly (L-lactic) acid (PLLA) in tetrahydrofuran (THF) in five major steps: (i) polymer dissolution, (ii) gelation, (iii) solvent extraction, (iv) freezing, and (v) freeze-drying (**Figure 3**) [24].

Similar to drawing, this method has a minimal equipment requirement, batch-to-batch consistency is achieved easily, and the mechanical properties of the matrix can be tailored by adjusting polymer concentration [14]. Nevertheless, it is limited to a few polymer types and thus cannot be scaled up and there is only little or no control over the diameters/dimensions of fibers formed.

1.1.4. Self-assembly

In self-assembly, smaller molecules are used as building blocks to create nanofibers [14]. The main mechanism for a generic self-assembly is the intermolecular forces that bring the smaller units together and the shape of the smaller units of molecules that determine the overall shape of the macromolecular nanofiber. The method has been used extensively in synthesis of genetic materials such as DNA [18, 19] and many other copolymers [19]. According to Hartgerink and his colleagues, during the synthesis of nanofibers based on this method, an individual, preexisting molecule (**Figure 4** left) is systematically arranged such that intermolecular forces can bind the concentrically arranged smaller units together (**Figure 4** middle). Thus, the small molecules organize themselves into the preferred pattern and function that results in the formation of macromolecular nanofiber mesh (**Figure 4** right) [19].

Researchers have extensively explored this method in producing fibers of diameters between 7 and 100 nm and fiber lengths between 1 and 20 micron using PCEMA core-PS shell in THF, PAA/y-Fe_2O_3 in THF, PS core-P4VP corona in chloroform and peptide-amphiphile in chloroform [18, 19]. Self-assembly requires no machinery to move or orient components. Self-assembly can be used to produce atomically precise nanosystems, meaning it is good for obtaining smaller nanofibers. It is simple, has easy processability, and is used to make one by one continuous nanofiber with uniform shape. However, it is a very complex process that cannot be scaled up, fiber dimensions cannot be controlled, its loading efficiency is poor, maintaining its porosity for longer duration poses serious challenge, and for every product, the structure of the parts must encode the structure of the whole [14].

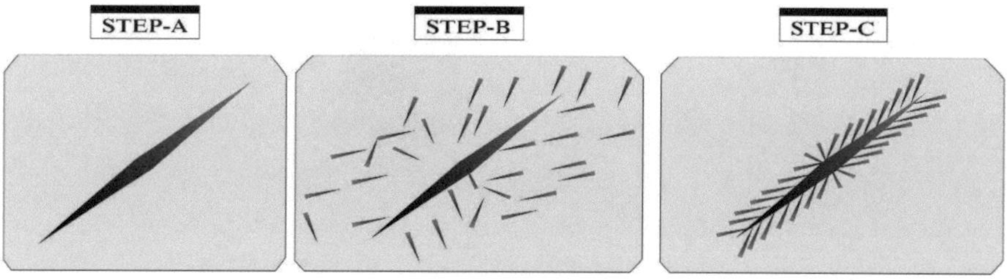

Figure 4. A typical representation of the self-assembly electrospinning process.

1.1.5. Melt blowing

Melt blowing involves the use of high-velocity air to produce fibers directly from polymers. It is a single-step process that converts polymer raw material directly into nanofibers. The polymer is melted in an extruder and then pumped through die holes into a high-speed, hot air chamber where the fibers formed are collected on a rotating collector (**Figure 5**) [20, 21]. The process has been widely used to generate nanofibers from polymers such as polypropylene, polyethylene, polybutylene terephthalate, Nylon 6, and polystyrene, and it has a high productivity [20, 21].

1.1.6. Electrospinning

The structure, morphology, and geometry of nanofibers and the porosity and tensile properties of nanofiber mats can be investigated through conventional techniques and instruments. Among these methods, electrospinning has been used to convert a large variety of polymers into nanofibers and has proven to be a process that has the potential for mass production. In simple terms, electrospinning involves the drawing of fluid, in the form of either molten polymer or polymer solution using electric field. In electrospinning, the needle attached to the syringe (containing the polymer solution/melt) is connected to high-voltage power source in order to create a high electric field between the polymer melt/solution and metallic fiber–collecting plate (**Figure 6**). The generation of sufficiently high electric field that can surmount the surface tension of the polymer solution results in the formation of droplets that travel toward the collector plate as a Taylor cone. Before the jet is collected on the aluminum foil plate as nanofiber mesh, the solvents dry off while the fiber jet is still traveling. The images in **Figure 6** show that the major electrospinning components are high-voltage power source, delivery channel for the viscoelastic polymer melt or solution, and collecting plates (be it flat or rotating plates) [12–14].

Whereas an external mechanical force drives the polymer solution/melt via a die in drawing method, electrospinning principle makes use of the high potential gradient generated from the power source connected to the needle to propel the droplet jet toward the collector. Upon

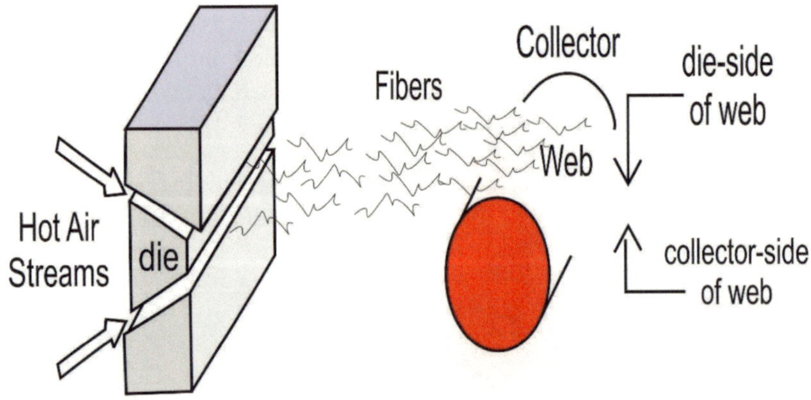

Figure 5. A typical melt blowing setup (Adapted from Ref. 21).

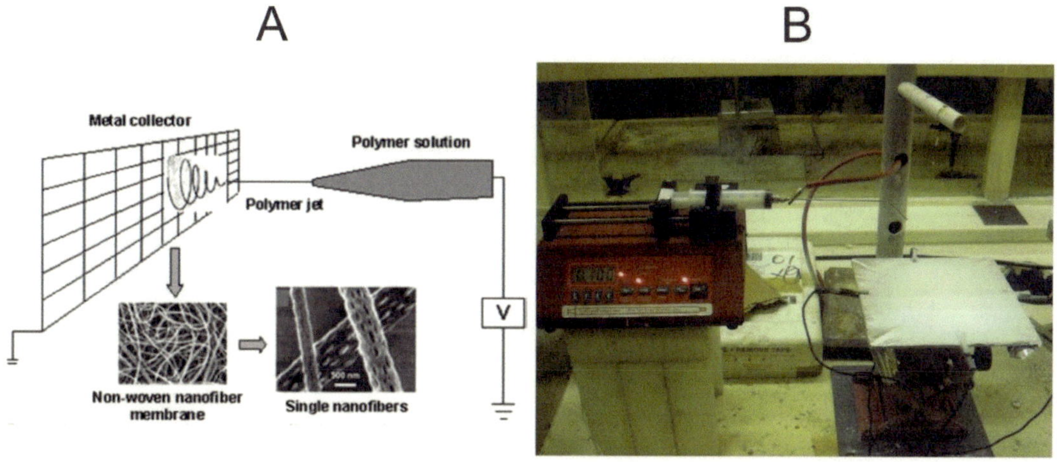

Figure 6. Schematic diagram to show polymer nanofibers by electrospinning (A) and the photograph of a typical electrospinning setup (B).

application of high voltage to the needle tip, surface charges accumulate and they lead to the deformation of the spherical droplet to a Taylor cone; beyond this point, a jet of polymer solution will therefore erupt from a polymer solution droplet. As the jet travels toward the collector, tensile forces brought about by surface charge repulsion lead to a bending motion. According to Taylor's theory, it is the instability induced on the surface of the electrically charged droplet that causes the nanofiber formation [25, 26]. Taylor hypothesized that a spherical droplet of polymer forms at the capillary tip and elongates as the applied voltage increases. The elongated droplet assumes a cone-like shape and a narrow jet of liquid ejects from this point [25, 26]. It is the change in shape of the droplet into conical shape that defines the onset of the fiber formation. The polymer chain entanglements within the solution will prevent the electrospinning jet from breaking up. The pendant electrically charged Taylor cone does not explode because of the chain entanglement in the concentrated polymeric solution. The surface area of the Taylor cone also increases to accommodate the charge buildup and this leads to stretching out of the cone and fiber formation [27].

Coupled with the ability of scaling up the electrospunnanofibers, one other unique advantage of electrospinning is that it is able to easily control the orientation of the nanofibers, which is important because fiber arrangement has a significant effect on the performance of the subsequent SPE sorbent devices [10–14]. Electrospinning technology is a useful, economical, and easily setup means of fabricating of 3D, highly porous, nanofibrous scaffolds tailored for a wide range of applications. Interestingly, several electrospun nanofibrous scaffolds have been revealed to support cellular activities and tissue formation [12–14]. The inherently high surface to volume ratio of electrospun scaffolds can enhance cell attachment, drug loading, and mass transfer properties. In addition to their high surface area to volume ratio, tunable porosity, and ability to manipulate nanofiber composition for desired properties, the nonwoven nanofibrous mats produced by electrospinning mimic extracellular matrix components much closely as compared to the conventional techniques [10–14]. Hence, the emergence of electrospinning/nanotechnology comes with a high potential for clinical applications. Until recently (5–10 years), electrospinning was not extensively employed as potential polymer-processing technique for applications in tissue engineering and drug delivery despite the long history of tissue engineering (about 30 years) [28]. The renewed interest can be ascribed to the aforementioned advantages of electrospinning over other conventional methods [28].

In electrospinning principle, the viscoelastic solution before transforming into nanofibers undergoes a number of processes. In 2006, Reneker and Fong divided the electrospinning process into four stages and they include launching the jet, jet elongation, whipping instability, and solidification [29].

1.1.6.1. Launching the jet

As the first stage of electrospinning process, the jet launching stage comprises droplet generation and Taylor cone formation. In the absence of an applied electric field, a viscoelastic solution pumped through a capillary will form normal droplets and fall off under the influence of gravity without forming fiber. It is the instability induced on the surface of the electrically charged droplet that causes the nanofiber formation [25, 26]. According to Deitzel et al., [30], the jet initiation occurs from the surface layers of the cone.

1.1.6.2. Jet elongation

In 2001, Buer revealed that the velocity of the jet increases as it travels toward the collector. It is believed that the potential difference between the collector and the point of release (needle tip) plays a major role in this process. As a result of solvent evaporation and polymer stretching, the jet diameter decreases rapidly. Resultantly, when a voltage (Ve) exceeding the strength of surface tension of the polymer solution is applied, jet elongation is expected to occur [29].

1.1.6.3. Whipping instability

As the straight fiber jet travels toward the collector plate, it tends to bend and displays undulating movements due to the competition between different forces/fields acting on the charged jet (axis symmetric, bending, and whipping and Raleigh instabilities). It is the interplay of all these forces that determines the diameter of the jet. However, whipping instability is the main mechanism responsible for reducing nanofiber dimensions because the predominant mode of instability exhibited is usually dependent on the applied electric field, and stronger fields favor whipping instability [31–33]. Whipping instability can be suppressed using a secondary electric field or a short gap distance (between the tip of the needle and the collector), but this does not affect the average fiber diameter significantly [33]. Due to the contribution of all these forces, there is not yet a mathematical model that has singly explained the entire electrospinning process.

1.1.6.4. Jet solidification

The solvent employed during electrospinning plays a major role in the solidification, morphology, mechanical integrity, and the microstructure of the electrospunnanofibers because the volatility characteristics of a chosen solvent is essential in the process and equally determines the time available to the jet to undergo whipping [34]. It has been pointed out that with appropriate selection of solvents and process parameters, extremely fine nanofibers can be electrospun [35].

1.1.6.5. Types of electrospinning

Attempts to improve the productivity of the electrospinning process have led to the development of three groups of improved and more efficient versions of electrospinning. Highlighted below are mononozzle, multinozzle, and needleless electrospinning.

1.1.6.5.1. Mononozzle electrospinning

The mononozzle is the simplest, cheapest, and most popular type of electrospinning setup in which only one nozzle/needle discharges the polymer solution with low productivity though (**Figure 7**).

1.1.6.5.2. Multinozzle electrospinning

Here, the polymer solution is fed into an array of nozzles or needles that allows for the deposition of multicomponent structures if different polymer solutions are electrospun concurrently, thereby increasing productivity of electrospinning by increasing the number of nozzles [36]. While the static type of needles remains immobile, the moving-type array of needles is programmed to move in unison (**Figure 8**). Fluctuations in electric field between the nozzles and the collector can influence the fiber diameters.

1.1.6.5.3. Needleless electrospinning

The needles are replaced with holes from where the nanofibers extrude with the application of optimal electrical charges induced on the viscoelastic solution by an electrode inside the porous tube. The needleless electrospinning comprise a porous polyethylene tube placed inside a coaxial cylindrical drum, and internal air pressure helps to push the polymer solution through the pores of the porous polyethylene tube (**Figure 9**). Although this is the most efficient electrospinning method, reproducibility is usually low due to the difficulty

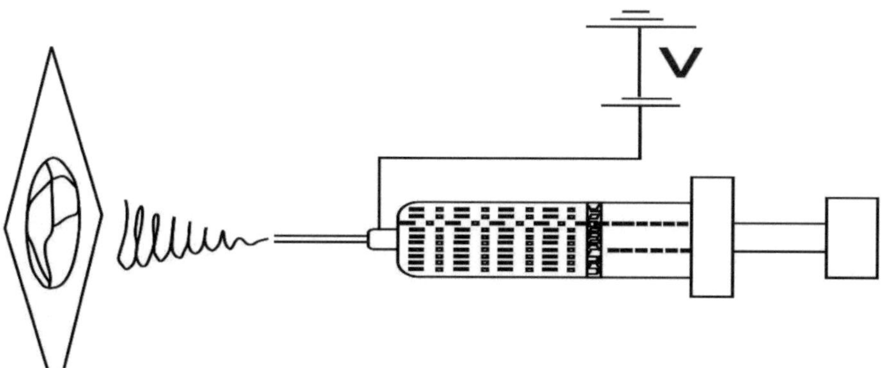

Figure 7. Schematic representation of a mononozzle electrospinning setup.

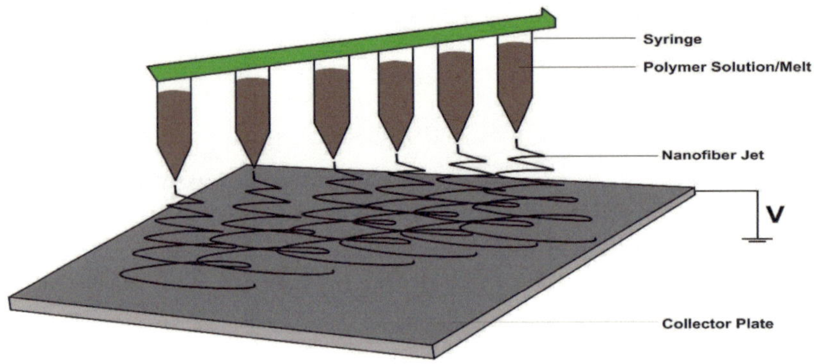

Figure 8. Schematic diagram of a multinozzle electrospinning setup.

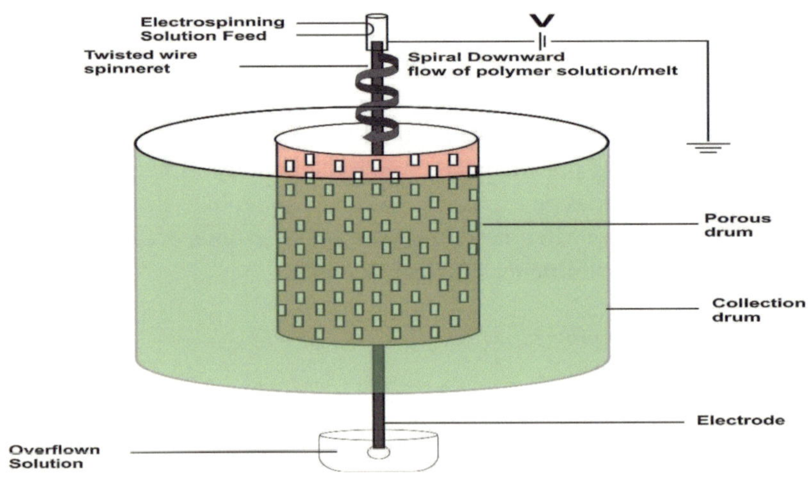

Figure 9. Schematic representation of a needleless electrospinning setup.

in maintaining an even distribution of polymer solution and air pressure over the different holes in the porous drum [37].

1.1.6.6. Processing parameters of electrospinning

The parameters affecting electrospinning and the fibers may be broadly classified into polymer solution parameters, processing conditions (applied voltage, temperature, and effect of collector), and ambient conditions. Understanding these parameters makes possible the fabrication of fibrous structures of various forms, morphology, and arrangements.

The processing conditions are important parameters that affect electrospinning process, and these are various external factors (applied voltage, the feed-rate, temperature of the solution, type of collector, diameter of needle, and distance between the needle tip and collector) acting on the electrospinning jet. Although processing conditions have significant influence on the fiber morphology, they are less significant than the solution parameters discussed earlier [10–14].

1.1.6.6.1. Applied voltage

A high voltage will induce the necessary charges on a solution and, together with the external electric field, will initiate the electrospinning process when the electrostatic force in the solution overcomes the surface tension of the solution. In most cases, a higher voltage will lead to greater stretching of the viscoelastic solution due to the greater columbic forces in the jet as well as the stronger electric field. These have the effect of reducing the diameter of the fibers [38, 39] and also encourage faster solvent evaporation to yield drier fibers [1, 2, 40]. However, researchers had correlated an increase in beads' density to increased voltage and suggested it may be the result of increased instability of the jet as the Taylor cone recedes into the syringe needle [30, 40]. This is because the greater amount of charges generated will cause the jet to accelerate faster and hence more volume of solution will be drawn from the tip of the needle without a corresponding increase from the source of supply. This may result in a smaller and less stable Taylor cone [41] or the Taylor cone may recede into the needle [30, 40]. Given the changes in the shape of beads from spindle-like to spherical-like with increasing voltage [41], Krishnappa et al. [42] reported that increasing voltage will increase the beads' density, which, at an even higher voltage, the beads will join to form a thicker diameter fiber.

On the contrary, when the voltage is low, the duo of weak electric field and reduced jet acceleration may surge the jet flight time, thereby resulting in finer fibers [43]. Based on this theoretical premise, finer fibers can be obtained using a near-critical voltage for electrospinning [43]. Generally, both high negative and positive voltage of more than 6 kV are able to cause the solution drop at the tip of the needle to distort into the shape of a Taylor cone during jet initiation [26].

In addition to affecting fiber physical appearance, the application of high voltage during electrospinning will likely increase the order of polymer molecules, resulting in higher polymer fiber crystallinity [1, 2, 10–14]. Nevertheless, polymer fiber crystallinity is reduced beyond a certain voltage [1, 2, 10–14].

1.1.6.6.2. Feed-rate of polymer solution

Feed rate is the quantity of polymer solution pumped into the tip per unit time. For a steady state and continuous formation of fibers, the feed rate of the solution must correspond to the rate of its removal from the tip. When the feed rate is increased, there is a corresponding increase in the stretching of the solution and fiber diameter since there is a greater volume of solution that is drawn away from the needle tip [44]. The large volume of solution drawn from the needle tip affords the solvents in the deposited fibers limited time to evaporate given the same flight time. The residual solvents may cause the fibers to fuse together where they make contact forming webs. Moreover, at higher feed rates, larger fiber diameters and beads often result [36]. A lower feed rate is highly desirable to give the solvent more time for evaporation [10–14]; however, the Taylor cone often gets depleted and the electrospinning process may only be intermittent or even stop completely.

1.1.6.6.3. Temperature of polymer solution

Having said that increased temperature decreases viscosity and surface tension, with a lower viscosity, the Columbic forces are able to exert a greater stretching force on the solution thus resulting in fibers of smaller and more uniform diameter [45]. Increased polymer molecule mobility due to increased temperature equally allows the Columbic force to stretch the solution further. Nevertheless, the use of high temperature in the electrospinning of biological composites (enzymes and proteins in polymer solution) is not advisable as it may cause the substance to lose its sensitive functionality.

1.1.6.6.4. Effect of collector

It is the electric field potential that is generated between the source (needle tip) and collector that initiates the release and deposition of fiber jets. The simplest and most commonly used collector is a stationary metal plate or an aluminum foil placed at a fixed distance from the needle tip. The collector plate is mostly made out of conductors that are electrically grounded so that there is a stable potential difference between the source and the collector to allow for a rapid discharge of the residual charges on the fibers.

In the case when a nonconductor is used as a collector, charges on the electrospinning jet will quickly accumulate on the collector, which will result in the deposit of fewer fibers with lower packing density (due to repulsive forces of accumulated charges) [14]. The material, nature, and geometry of the collector play a major role in defining the morphology of the fibers [13]. Several different shapes of collectors such as flat plate; rotating drum, mandrel, rotating disc, rectangular, triangular, or wire cylinder frame; electrode pair arrangements; ring and mesh electrode; and cones have been reported [1, 2, 10–14, 26]. It has also been demonstrated that some of the common solvents such as water [46] and methanol [46] could be better collectors than their solid counterparts. A liquid collector may equally be used to precipitate the nanofibers when nonvolatile solvents are used. Srinivasan and Reneker demonstrated this by electrospinning poly (p-phenylene terephthalamide) nanofibers from sulfuric acid solution into a grounded water bath to precipitate the polymer [47].

The nature of the collector (static or moving) significantly affects the electrospinning process. While rotating collector has been used to collect well-aligned fibers, it was found to assist in yielding fibers that are dry [48]. Hence, it is very useful for certain solvents like dimethylformamide (DMF), which is good for electrospinning but has a high boiling point that may result in the fibers being wet when they are collected. A rotating collector will give the solvent more time to evaporate [48] and also increase the rate of evaporation of the solvents on the fibers. This will certainly improve the morphology of the fiber especially where distinct fibers are required.

Studies have shown that the porosity of the collector plate affects the packing density of the deposited fibers [10–14]. Fiber meshes of lower packing density were collected on porous collectors (paper and metal mesh) as compared to the smooth surfaces such as metal foils [10–14]. Responsible for this observation were higher evaporation rate and faster diffusion of the remaining solvents on the fibers collected on the paper and metal mesh (due to higher surface area) as compared to the smooth surfaces that may cause an accumulation of solvents around the fibers due to slower evaporation rate [10–14]. The residual solvents on these fibers pull them together to give a more densely packed structure unlike the dried fibers on the porous collector where the residual charges remaining on the fiber repel subsequent fibers [10–14]. Nevertheless, on a smooth surface, the residual solvents will encourage the residual charges to be conducted away from the collector.

1.1.6.6.5. Distance between tip and collector

This is the distance extending from the capillary tip to the surface of the collector. The gap distance has a direct influence on both the flight time (available for solvent evaporation) and the electric field strength [1, 2, 13]. For independent fibers to form, the electrospinning jet must be allowed time for most of the solvents to evaporate; otherwise, shortening the gap distance implies that the excess solvent may cause the fibers to merge where they contact to form junctions resulting in inter- and intralayer bonding of interconnected fiber mesh, which may provide additional strength to the resultant scaffold [13]. Depending on the solution property, the effect of varying the distance may or may not have a significant effect on the fiber morphology. In some cases, changing the distance has no significant effect on the fiber diameter, but Megelski and his coworkers observed bead formation when distance was too low and it was attributed to increased field strength (high voltage) [39]. Having said that the field strength is too high, the increased instability of the jet may encourage bead formation [40], and increasing the distance results in a decrease in the average fiber diameter because the longer distance means that there is a longer flight time for the solution to be stretched before it is deposited on the collector [29, 43]. However, Zhao et al. [43] reported a case where no fibers were deposited on the collector due to very large distance and Lee et al. [49] revealed at a longer distance, and the fiber diameter increased due to the decrease in the electrostatic field strength resulting in less stretching of the fibers Lee et al. [49]. Hence, it suggests that there is an optimal electrostatic field strength below which the stretching of the solution will decrease resulting in increased fiber diameters [13, 14].

1.1.6.6.6. Diameter of pipette orifice/needle

Conducting materials such as metal needles as well as nonconducting materials such as glass and plastics have been used as the capillary tip in solution electrospinning [50], and their internal diameter or pipette orifice has a significant effect on the electrospinning process. A smaller internal diameter was found to reduce the fiber diameter, clogging, as well as the amount of beads on the electrospun fibers [43]. When the size of the droplet at the tip of the orifice is decreased, surface tension of the droplet increases, a greater columbic force is therefore required to cause jet initiation for the same voltage applied [43]. Consequently, the acceleration of the jet decreases and this allows more time for the solution to be stretched and elongated before it is collected. Also, the reduction in clogging could be due to less exposure of the solution to the atmosphere during electrospinning. However, if the diameter of the orifice is too small, it may not be possible to extrude a droplet of solution at the tip of the orifice [43].

Although most studies have reported the use of a simple and static capillary tip, a number of innovations have explored the use of movable tips. Kidoaki et al. used a movable tip to obtain an even deposition of nanofibers on a drum collector, and the moving tip helps to align the fibers evenly [36]. Li also used a tip made of a nonconducting fiber inserted in the lumen of a conducting capillary tip [51]. Ultimately, this modified tip allowed the electric field to be used solely to accelerate the jet and therefore reduced the potential needed to be applied [51].

1.1.6.6.7. Polymer solution parameters

The surface tension and viscosity of polymer solutions or melt significantly affect the morphology of fibers obtained from electrospinning. While surface tension contributes to the formation of beads, viscosity and electrical potential largely determine the extent of elongation and diameter of the electrospun fibers [14].

1.1.6.6.8. Molecular weight and solution viscosity

Viscosity is a measure of the resistance of a material to flow. Viscosity is a prerequisite for a successful electrospinning because the polymer solution or melt must be viscoelastic in order to stretch the electrically driven fiber jet from the needle tip toward the collector plate without breaking up. It is the entanglement of the molecule chains that is often determined by the length of polymer chain that prevents the fiber jet from breaking up, thus maintaining a continuous solution jet. Consequently, monomeric polymer solution does not form fibers when electrospun [13, 14].

It is worth noting that the viscosity of a polymer solution is directly affected by its molecular weight (length of polymer chain) since the polymer length determines the extent of entanglement of the polymer chains in a given solvent. Hence, employing polymers with high molecular weight increases viscosity just as increasing the polymer concentration equally increases viscosity. When viscosity is too low, surface tension overrides and hence electrospraying occurs or beaded fibers are formed, whereas with increased viscosity, the diameter of the fiber also increases probably due to the greater resistance of the solution to be stretched by the charges on the [39, 40, 48].

1.1.6.6.9. Surface tension and temperature

Surface tension is that force that is acting on the surface of a solution to hold it in place. It is the main force of attraction that opposes the coulomb repulsion in electrospinning. When a very small drop of water falls through the air, the droplet generally takes up a spherical shape due to its surface tension. Surface tension usually decreases the surface area per unit mass of a fluid. The initiation of electrospinning requires the charged solution to overcome its surface tension. However, as the jet travels toward the collection plate, the charged solution is stretched, while the surface tension may cause the formation of beads along the jet or breakup of solutions into droplets [13, 14]. To encourage the formation of smooth fibers, surfactants are added to solutions to reduce surface tension or addition of solvents with low surface tension (ethanol) [13, 14].

At molecular level, liquid molecules at high temperature gain kinetic energy that increases their collision speed during their Brownian movement. Consequently, the molecules in cooler liquid bind more strongly than the loosely bound rapid-moving molecules [12–14]. Hence, surface tension drops when temperature is increased owing to decrease in bonding energy between the molecules [12–14].

1.1.6.6.10. Conductivity of solution and electric charge (voltage)

To initiate the electrospinning process, the repulsive forces within the solution must surmount the solution surface tension. The repulsive forces are as a result of the sufficient charges acquired by the solution molecules. These charges that the solution carries often play crucial role in the subsequent stretching or drawing of the electrospinning jet. Having said that electrospinning involves the stretching of solution caused by repulsion of the charges at its surface, an increase in the conductivity means more charges can be carried by the electrospinning jet. Since the presence of ions increases the conductivity of the solution, the critical voltage for electrospinning to occur will equally reduce [12–14]. The increased charges also result in a greater bending instability. As a result, the deposition area of the fibers is increased [12–14]. This will also favor the formation of finer fibers since the jet path is now increased.

To increase the conductivity of the solution at the same time reducing the surface tension, ionic surfactant such as triethyl benzyl ammonium chloride is added although it often reduces the fiber diameter [12–14]. However, there is a limit to the reduction in the fiber diameter because in some cases, the addition of ionic salt may cause an increase in the viscosity of the solution improving conductivity; thus, the viscoelastic force is stronger than the columbic forces of the charges resulting in an increased fiber diameter instead [45]. Carboxylic acids, mineral acids, some complexes of acids with amine, mineral salts, some tetraalkylammonium salts, and stannous chloride, among others can also increase solvent conductivity when introduced [13]. For example, the addition of a small amount of water to organic acid solvents will greatly increase their conductivity due to ionization of the solvent molecules. pH adjustment is another way of increasing conductivity. Mixing of chemically noninteracting components will increase conductivity as well [13, 14].

1.1.6.6.11. Dielectric effect of solvent

The dielectric constant (Ɛ) is a measure of how effectively a material placed in electric field can concentrate the electrostatic lines of flux, that is, the solvent's ability to hold electrical charges. It has been widely reported that solvents or solutions with high dielectric properties such as N,N-dimethylformamide (DMF) enhance fiber morphology by reducing fiber diameter and bead formation [1, 2, 48, 52]. This is feasible because solutions with higher dielectric constants tend to disperse the surface charge density on the jet more evenly and this leads to the production of fibers with uniform morphologies and smaller diameters [48].

In a study by Min and his coworkers, they compared the morphologies of nanofibers electrospun from 15 wt% poly (lactide-co-glycolide) solutions in chloroform and in hexafluoropropylene (HFP) [53]. Results showed that different fiber morphologies were obtained with the two solvents and the average fiber diameter obtained from HFP (having a higher Ɛ of about 16.7) was lower than those obtained from chloroform (having a lower value Ɛ of about 4.81). Son et al. corroborated this fact when they spun polyethylene oxide (PEO) in different solvents where the solvents with higher Ɛ resulted in smaller average diameters [52]. The interaction between an electrospinning solution and any solvent that increases the solution's dielectric constant will equally affect fiber morphology. This implies that only the high dielectric constant of the solvent introduced into a polymer solution is not the only factor that impacts the fiber morphology: the interaction between the mixtures also impacts the resulting fiber morphology. Bead formation was reported despite introducing DMF (high dielectric constant solvent) into polystyrene solution to improve electrospinnability and fiber morphology [48]. The investigators posit that the poor interaction between the polystyrene solution and DMF could possibly explain the beaded fibers formed in place of finer fibers [48] (**Table 1**).

Solvent	Chemical structure	Conductivity (S.m^{-1})	Viscosity (mPa.S)	Surface tension (mN.m^{-1})	Dielectric constant (F/ms or cP)
Hexafluoroisopropanol (HFIP)	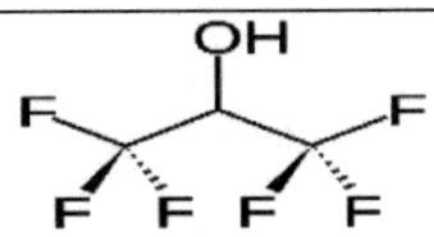	—	1.25	16.1	16.7
Formic acid (FA)	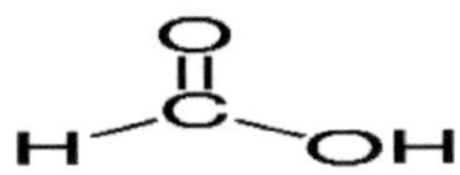	6.4×10^{-7}	1.96	37	58
Dichloroethane	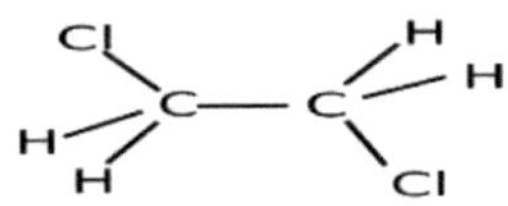	3.0×10^{-8}	0.84	32.23	10.45
2-Propanol	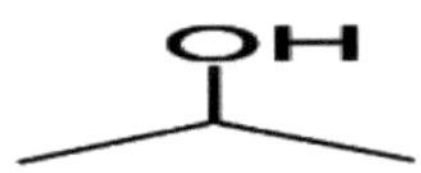	6.0×10^{-6}	2.4	23.3	18.6
Acetic acid	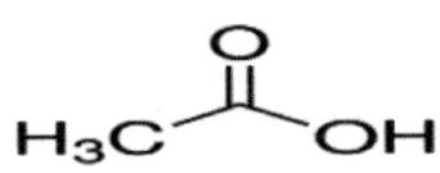	6×10^{-9}	1.12	26.9	6.15
Acetone		2×10^{-7}	0.32	24	20.7
Acetonitrile	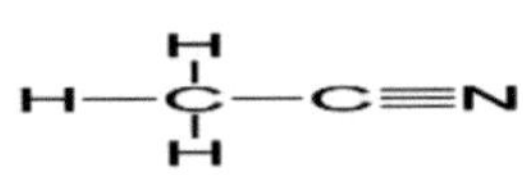	7×10^{-6}	0.352	29.29	36
Chloroform	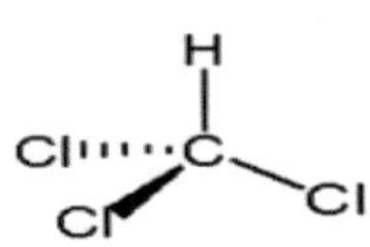	$<10^{-10}$	0.56	27.14	4.8–4.9
Dichloromethane	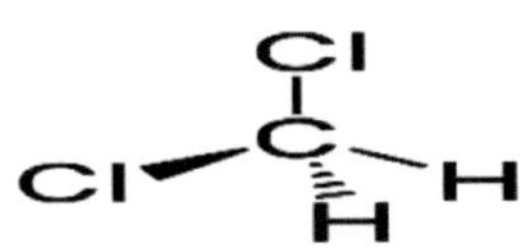	4.3×10^{-11}	0.42	28.1	9.1
Dimethylformamide	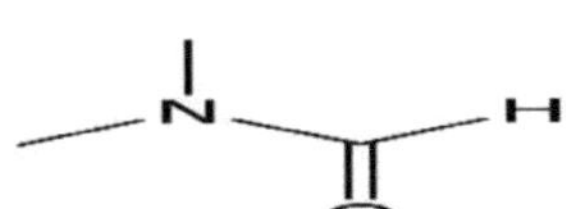	6×10^{-8}	0.8	35.2	36.71
Ethyl acetate	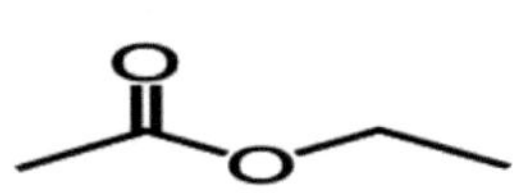	2×10^{-7}	0.46	23.2	6.0
Ethanol		1.3×10^{-7}	1.04	22.39	24.55
m-Cresol	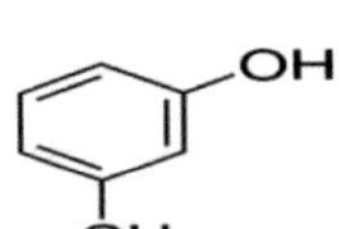	—	0.45	41.7	11.8
Methanol	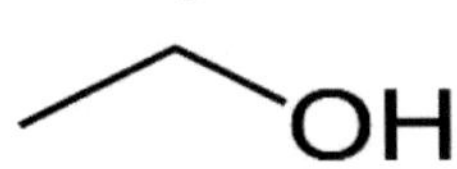	3.0×10^{-7}	0.544	22.07	32.6

Solvent	Chemical structure	Conductivity (S.m^{-1})	Viscosity (mPa.S)	Surface tension (mN.m^{-1})	Dielectric constant (F/ms or cP)
Pyridine	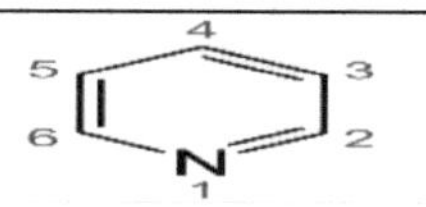	—	0.974	38	12.3
Tetrahydrofuran	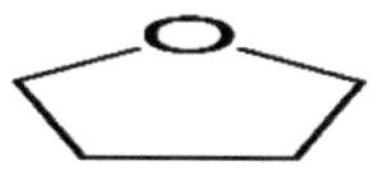	1.5×10^{-11}	0.48	28.4	7.6
Toluene	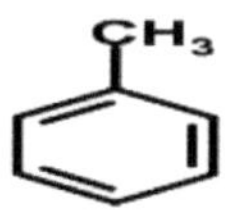	4.0×10^{-10}	0.59	27.6	2.438
Trifluoroethanol		—	1.64	19.4	27
Water	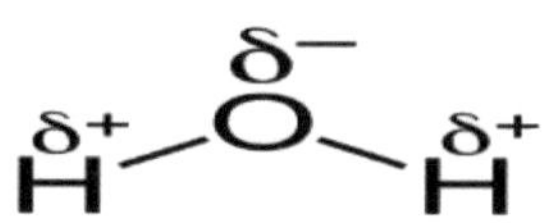	5.5×10^{-6}	0.890	72.75	80.2

Table 1. Dielectric constant of common electrospinning solvents.

1.1.6.6.12. Ambient parameters

Since electrospinning is driven by external electric field, changes in its surrounding are expected to affect the process. While high humidity has been found to cause the formation of pores on the fiber surfaces, other environmental factors like pressure, temperature, and type of atmosphere have equally been investigated [1, 2].

1.1.6.6.13. Humidity

Humidity has been revealed to affect fiber morphology especially if it is spun from a volatile solvent [39] because at high humidity water condenses on the surface of fibers and affects the rate of evaporation of solvents. For example, Casper and colleagues showed that smooth fibers are formed from PSU dissolved in THF only when humidity is less than 50%; circular pores were formed on the fiber surfaces when humidity was higher than 50% with their pore depth and sizes increasing with increased humidity [54].

Low humidity aids volatile solvents to dry up faster. However, if the rate of solvent escape/removal from needle tip is slower than evaporation rate of the solvent, there is a good chance of needle clogging during electrospinning. It has been hypothesized that high humidity can help the discharge of electrostatic charges on electrospun fiber [51], but with decreasing humidity, there was an increase in the amount of charge on the particle. This is evidenced by a study on glass particles transported in a grounded copper pipe, where no charges were found on the particles at higher relative humidity (>76%) [51].

1.1.6.6.14. Type of atmosphere

Different gases behave differently under high electrostatic field. While helium breaks down under high electrostatic field rendering electrospinning impossible, Freon®-12 with higher breakdown voltage yields fibers with twice the diameter of those electrospun in air given all other conditions equal. For example, using a positively charged capillary tip in an electron-rich gaseous environment will impede the process, and an environment of highly electronegative gases (such as CO_2 or Freons) discourages the loss of surface charges and improves nanofiber quality [55].

1.1.6.6.15. Pressure

It is only possible to investigate the effect of pressure on electrospinning jet under an enclosed condition. Electrospinning is not possible at very low pressures due to direct discharge of the electrical charges [14]. The fluidity of a polymer melt/solution is rapidly increased below atmospheric pressure as the melt/solution bubbles rapidly at the needle tip, often causing unstable jet initiation. However, at high pressure, air may be delivered coaxially to the needle tip to provide an additional drag force for jet extension. When this drag force is dominant over the electrostatic force in driving jet extension, the process is referred to as electroblowing [13, 14].

1.1.6.6.16. Temperature

Given that temperature reduces viscosity and determines the rate of evaporation of the solvents in fiber jet, it controls the final diameter of the nanofibers. Um et al. used a jacket of heated air (25–57°C) to decrease solution viscosity and increase the rate of drying of the fiber when electrospinning an aqueous solution of hyaluronic acid (HA) [56]. Subramanian et al. suggested the use of an external heating source such as a heat gun or a high wattage lamp as an alternative for drying fibers [57].

1.2. Some notable pharmaceutical and medical applications of electrospunnanofibers

Electrospunnanofibers have found promising applications in various biomedical areas. Almost all of the human tissues and organs are deposited in nanofibrous forms or structures, such as skin, bone, dentin, collagen, and cartilage to mention but a few. Consequently, business opportunities for nanostructured materials in biomedical applications are estimated to be of the order of 180 billion US dollars in 2015 [12–14]. A few applications of ENs are highlighted below.

1.2.1. Electrospun polymer nanofibers as a solid-phase extraction sorbent

The amazing properties of polymer nanofibers have rekindled their interest in several vital applications. They have larger surface area to volume ratio (can be 103 times of that of a microfiber), ease of surface functionalities, and better mechanical performance (e.g., stiffness and tensile strength) as compared to their microfiber counterpart or any other known form of the polymer. [1, 2, 10, 11].

It has been demonstrated that the fiber-packed SPE device offers a reduced pressure drop during the extraction and the desorption compared to a conventional particle-packed SPE cartridge [1, 2, 10, 11]. Electrospunnanofibers exhibit superb analytical potential as they can be miniaturized into nanoscale to simplify sample preparation, preconcentration owing to their effective interaction with samples, and the ease of incorporating different chemistries [1, 2]. These SPE sorbents with high miniaturization tendencies, fast extraction, less solvent, cost-effective, excellent analyte recoveries, good sorptive capacity, high selectivity, and good mechanical and chemical stability tend to satisfy all the current sample preparation requirement; hence, it is the choice sample preparation protocol for several studies [1, 2].

Their large surface area makes them equally applicable as wonderful sensing tools as applied in test strips and other colorimetric probes. Hence, significant effort has been made to increase the surface area of the sensing interface in chemical sensors [58].

1.2.2. Electrospunnanofibers composite in colorimetric probes (medical diagnostic tool)

As casualties from cancer mortality continues to grow due to the absence of state-of-the-art health facilities, there have been increasing demands for simple, rapid, highly sensitive, inexpensive yet reliable method for the early detection of cancer susceptibility. To this end, equipment-free sensor systems such as colorimetric detection by naked eye would be among the best and most practically useful methods as this could easily be fabricated into a test strip. Colorimetric biosensing does not require expensive or sophisticated instrumentation and maybe applied to field analysis and point of care diagnosis since color changes can be read by naked eye [58].

Nanoparticles often have unexpected visible properties because they are small enough to confine their electrons and produce quantum confinement effect: Their high surface activity and freely moving electrons could bring about Raman-enhanced light scattering effect and agglomeration. The high surface area to volume ratio reduces their incipient melting temperature. They often possess high molar extinction values, which makes them far more sensitive and stable than other conventional colorimetric probes [58]. Surface plasmon resonance (SPR) is a conspicuous property of metal nanoparticles that results from the combined resonance oscillations of the free electrons of the conduction band of the metal [58]. A sharp and intense absorption band is observed in the visible range [58]. In addition, electrospunnanofibers offer high surface area and porous membrane structure that are applicable for sensitive and fast sensing, which have shown improved sensitivities over conventional materials for applications such as gas sensors, chemical sensors, optical sensors, and biosensors [58].

In the recent past, there has been great interest in the unique mechanical, electrical, chemical, and optical properties that can be achieved by combining the advantages of gold nanoparticles and polymer nanofibers [58]. During the formation of nanocomposites, there is a high aggregation tendencies in the polymer matrix since they are metastable compared to their bulk materials due to the positive excess interfacial free energy [58]. Owing to the fact that their spatial distribution is the key to their optical property, there is a burning desire by analytical and pharmaceutical chemists to attain good dispersion during synthesis. Hence, the stabilization of nanoparticles in polymer matrix is a fundamental prerequisite in nanoscience technology. For example, the *in situ* one-step fabrication of gold nanoparticle filled-polymer nanofiber for the colorimetric probe of the carcinogenic biomarker 1-hydroxypyrene was demonstrated [58]. The fibers were fabricated in the form of a test strip for the biological monitoring of carcinogenic PAH exposure [58].

1.2.3. Electrospunnanofibers in drug release control

Controlled release is an efficient process of delivering drugs in medical therapy. This process is able to balance the delivery kinetics, minimize toxicity and side effects, and improve patient convenience. In a controlled release system, the active substance is loaded into a carrier or device first and then released at a predictable rate *in vivo* when administered by an injected or noninjected route. As a potential drug delivery carrier, electrospunnanofibers have exhibited many advantages. The drug loading is very easy to implement via electrospinning process, and the high applies voltage used in the electrospinning process had little influence on the drug activity. The high specific surface area and short diffusion passage length give the nanofiber drug system a higher overall release rate than the bulk material (e.g., film), while the release profile can be finely controlled by modulation of nanofiber morphology, porosity, and composition. Coating of the nanofibers with a polymer shell could also be an effective way of controlling the drug release profile. The early burst release of drugs can be lowered by encapsulating water-soluble drugs into nanoparticles, followed by incorporating the drug-loaded nanoparticles into nanofibers. Also, the rate of releasing a water-soluble drug could be slowed down when nanofiber matrices are cross-linked [59]. Nanofibers for drug release systems are mainly made by electrospinning of biodegradable polymers, such as polylactic acid (PLA), poly(D-lactide) (PDLA), poly(L-lactide) (PLLA), poly(lactic-co-glycolic acid (PLGA), and hydrophilic polymers such as polyvinyl alcohol (PVA), polyethylene glycol (PEG), and polyethylene oxide. The release of macromolecules, such as DNA and bioactive proteins from nanofibers, has also been investigated [59]. The main factors influencing the release performance of drugs include the type of polymers used, hydrophilicity and hydrophobicity of drugs and polymers, solubility, drug-polymer comparability, additives, and the existence of enzymes in the buffer solution.

The need for the control release of water-soluble drugs usually associated with an early-stage burst cannot be overemphasized. Although for applications like preventing postsurgery-induced adhesion, early-burst release will be an ideal profile because most infections occur within the first few hours after surgery, for a long-lasting release process, it would be crucial to maintain an even and stable release. This is usually achieved by encapsulating the drug within a nanofiber matrix.

Alternatively, for water-insoluble drugs, the drug release from hydrophobic nanofibers into buffer solution is often difficult. In order to release the drug at a constant rate, an enzyme capable of degrading nanofibers is often mixed with the buffer. For example, Kim and his coworkers showed that rifampicin encapsulated in PLA nanofibers did release any drug, while the same polymer/drug with potassium revealed a zero kinetics drug-release profile [59].

1.2.4. Environmental remediation

Environmental quality issues are complex, challenging, and ever expanding; hence, regulatory bodies are increasing the amount of environmental monitoring required to ensure public safety. The high porosity, interconnectivity, microscale interstitial space, and large surface-to-volume ratio of nonwoven electrospunnanofiber meshes make them an excellent size-exclusion membrane for particulate removal from wastewater and air. The electrospunnanofiber membranes can effectively and successfully remove particles 3–10 μm in size (>95% rejection) from aqueous solution without a significant drop in flux performance and airborne particles with diameters between 1 and 5 μm by both physical trapping and adsorption (~100% rejection). In the environmental industry, affinity membranes have found applications in organic waste removal and heavy metal removal in water treatment. Affinity membranes are a broad class of membranes that selectively capture specific target molecules (or ligates) by immobilizing a specific capturing agent (or ligand) onto the membrane surface [12]. The use of enzymes like laccase or tyrosinase for the degradation of water pollutants such as endocrine disruptors has equally been reported: enzymatic remediation of endocrine disruptors in water [60].

1.2.5. Application of electrospunnanofibers in tissue engineering and scaffolding material in cell culture

For the treatment of tissues or organs that malfunction in a human body, one of the challenges to the field of tissue engineering/biomaterials is the design of ideal scaffolds/synthetic matrices that can perfectly mimic the structure and biological functions of the natural extracellular matrix (ECM). Of particular interest in tissue engineering is the fabrication of nontoxic, reproducible, and biocompatible three-dimensional scaffolds with high porosity of evenly distributed pore sizes, high surface area, structural integrity, biodegradability with the degradation rate matching the rate of neotissue formation that can positively interact with human cells to promote cell adhesion, proliferation, migration, and differentiated cell function for various tissue repair and replacement procedures. Recently, scientists have started paying much attention to making such scaffolds with synthetic biopolymers and/or biodegradable electrospun polymer nanofibers as they are believed to possess all the aforementioned qualities and can mimic the human native ECMs [1, 2]. Having said that human cells can attach and organize well around fibers with diameters smaller than those of the cells, nanoscale fibrous scaffolds can provide an optimal template for cells to seed, migrate, and grow.

Although the mechanisms by which a nanofibrous scaffold acts as a selective substrate are not yet known, a number of researchers have shown that the enhanced adsorption of cell adhesion matrix molecules enhances cell adhesion [61, 62]. Nikolovski et al. [61] and Woo et al. [62] have reported that electrospun nanofibrous scaffolds exhibited enhanced adsorption of cell adhesion ECM molecules, which may therefore enhance cell adhesion. Although fibronectin and vitronectin preferentially adsorbed to the nanofibrous scaffold at a level that was 2–4 times higher than those adsorbed to the solid-walled scaffold [62], vitronectin, rather than fibronectin, was the predominant matrix protein adsorbed from serum-containing medium onto poly(glycolic acid) (PGA) and poly(lactic acid) (PLA) [61].

While cells synthesize, assemble, organize, and maintain ECM macromolecules, ECM in itself provides structural integrity to the resident cells and acts as a messenger regulating cellular activities [63]. It is therefore expected that a scaffold that serves as a functional, temporary ECM must involve optimal cell-matrix interactions, as well as cell-cell interactions. Li and his coworkers evaluated the influence of the structural properties of biomaterial scaffolds on the biological activities of chondrocytes cultured in microfiber- and nanofiber-based scaffolds [8, 64]. While chondrocytes seeded onto nanofibrous scaffolds maintain a chondrocyte-like morphology, chondrocytes seeded into microfibrous scaffolds display dedifferentiated, fibroblast-like morphology [8, 64]. Yang and colleagues also demonstrated that a greater percentage of neural stem cells cultured on nanofibrous scaffolds exhibit a neuron-like morphology with longer neurite outgrowth as compared to their microfibrous scaffold counterpart [65]. Nanoscale fibers are believed to be smaller than cells by two orders and provide 3D environment that better promotes cell-cell and cell-ECM interaction, unlike the microfibers [64].

Studies have equally shown that nanofibrous scaffolds support multidifferentiation of mesenchymal stem cells (MSCs) [64]. With the aid of gene expression analysis and immunohistochemical detection of lineage-specific marker molecules, Li's group confirmed the formation of nanofibrous constructs containing mesenchymal stem cell (MSC) differentiated along adipogenic, chondrogenic, or osteogenic lineages [64]. It has been reported that basement membrane matrix (feeder cell layers) of embryonic fibroblast provides chemical and physical cue on embryonic stem cells (ESCs) and regulate their ability to self-renew and differentiate [66]. Hence, ESCs had successfully been cultured on nanofibrous scaffold resembling a basement membrane matrix, yet their "stemness" properties were reportedly maintained [67]. Given that nanofibrous scaffolds act as synthetic ECM network, providing physical and chemical cues to cells via cell-ECM interaction, nanofibrous scaffolds promote in vivo–like 3D matrix adhesion and activate cell signaling pathway. For example, Schindler and coworkers have demonstrated that nanofibrous cultures promote in vivo–like cell morphology of both fibroblasts and kidney cells [68].

1.2.6. Electrospunnanofibers as wound dressing material

Nonwoven nanofibrous membrane mats for wound dressing generally possess highly porous, well-interconnected pore sizes ranging from 500 to 1 mm, small enough to protect the wound from bacterial penetration via aerosol particle capturing mechanisms while exuding fluid from the wound [1, 2, 12–14]. Their high surface areas of 5–100 m^2/g are extremely efficient for fluid absorption and dermal delivery and generally assist the control of fluid drainage. Polymer nanofibers can also be used for the treatment of wounds or burns of a human skin, as well as designed for hemostatic devices with some unique characteristics. Through electrospraying or electrospinning, fine fibers of biodegradable polymers can be directly sprayed/spun onto the injured location of skin to form a fibrous mat dressing (**Figure 10**), which can let wounds heal by encouraging the formation of normal skin growth and eliminate the formation of scar tissue, which is usually a problem with traditional treatment [12–14]. Electrospinning also offers a simple way to add drugs into the nanofibers for possible medical treatment and antibacterial purposes [1, 2, 11]. For example, studies have shown that the successful incorporation of antibiotic agents into ENs prevents postsurgery abdominal adhesions and improves

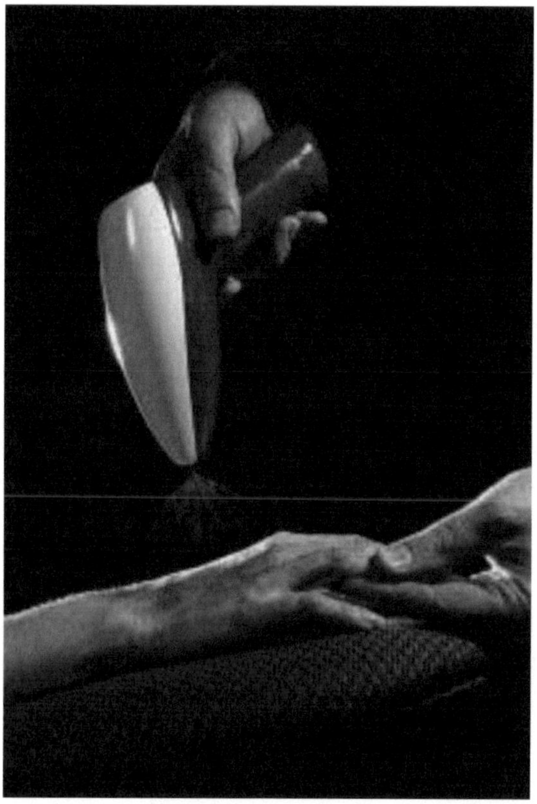

Figure 10. Nanofibers for wound dressing (www.electrosols.com).

the healing process [13, 14, 69]. Given their higher loading efficiency, superior mechanical performance (stiffness and tensile strength), controlled release ability, and excellent stability, ENs help in the delivery of plasmid DNA, large protein drugs, genetic materials, and autologous stem cell to the target site [69].

1.2.7. Electrospunnanofibers for immobilization of digestive enzymes

Chemical reactions using enzymes as catalysis have high selectivity and mild reaction conditions. For easy separation from the reaction solution, enzymes are normally immobilized with a carrier. The immobilization efficiently mainly depends on the porous structure and enzyme-matrix interaction. Nanostructured materials were recently used as enzyme carriers because of their large specific surface area and the high loading capacity. Many approaches have been used to immobilize enzymes on electrospunnanofibers including grafting enzymes on fiber surface, physical adsorption, and incorporating of enzyme into nanofiber via electrospinning followed by cross-linking reaction. To graft enzymes on nanofiber surface, the polymer used should possess reactive groups for chemical bonding. The immobilized enzymes normally showed a slightly reduced activity in aqueous environment compared with the unimmobilized native counterpart, although the activity in nonaqueous solution was much higher. For example, α-chymotrypsin was used as a model enzyme to bond chemically on the surface of electrospun PS nanofibers. The hydrolytic activity of the enzyme loaded was found to be 65% of the native enzyme, while the activity in nonaqueous solution was over 3 orders of magnitude higher than that of its native enzymes under the same conditions [70]. In another study that employed polyacrylonitrile nanofibers to immobilize lipase, the tensile strength of the nanofiber membrane was improved after lipase immobilization, and the immobilized lipase retained >90% of its initial reactivity. Also the immobilized lipase still retained 70% of its specific activity after 10 repeated reaction cycles with improved pH and thermal stabilities [71, 72].

When redoxases were immobilized, the incorporation of carbon nanotubes into nanofibers apparently increased the enzyme uptake and the activity of the immobilized enzyme was enhanced. The presence of carbon nanotubes also improved the stability of the redoxases immobilized. This improvement in catalysis performance was attributed to the fact that carbon nanotubes could behave as electron transferors to donate/accept electrons during enzyme catalysis or render the composite nanofibers higher biocompatibility. In certain studies, enzymes have been incorporated into nanofibers via electrospinning and subsequent cross-linking of the incorporated enzymes effectively prevented their leaching. For instance, in the presence of PEO or PVA, casein and lipase were electrospun into ultra-thin fibers, on cross-linking with 4,4-methylenebis (phenyl isocyanate), the fibers became insoluble, and the lipase encapsulated exhibited six times higher hydrolytic activity [73].

1.2.8. Electrospunnanofibers for catalyst carriers

In medicine, a carrier for catalyst is used to preserve high catalysis activity, increase the stability and life of the catalyst, and simplify reaction processes. Electrospunnanofiber mats are used as catalyst carrier, because their extremely large surface area could provide a huge number of active sites, thus enhancing the catalytic activity. The well-interconnected small pores in the nanofiber mat warrant effective interactions between the reactant and catalyst, which is valuable for continuous-flow chemical reactions or biological processes. Also, the catalyst can be grafted onto the electrospunnanofiber surface via surface coating or surface modification [74, 75]. For instance, Pd-loaded PAN-acrylic acid nanofibers were confirmed to have high activity and good recycling property for hydrogenation of an olefin at room temperature. The yield of hexene to hexane catalyzed by the palladium /PAN-AA nanofibers was 4.7 times higher than that of Pd/γ-Al_2O_3 [74]. In addition to Pd nanoparticles, Ag nanoparticles were incorporated into silica nanofibers, and the hybrid Ag-silica nanofibers sowed catalytic activity to assist $NaBH_4$ to reduce decomposition of methylene blue [44]. Photocatalysts, such a titania (TiO_2) and TiO_2-SiO_2, were also electrospun into nanofibers, and the photocatalytic activity was evaluated. In comparison to other nanostructured TiO_2 materials, such as commercial TiO_2 nanoparticles and mesoporous TiO_2, the nanofibers exhibited higher photocatalytic activities toward the degradation of methylene blue and gaseous formaldehyde [76].

1.2.9. Electrospunnanofibers as chemical sensors for carcinogenic heavy metals in medical diagnosis

The challenge in the field of new approaches toward high sensitivity detection techniques remains a major challenge in the field of chemical sensing. Sensors have been widely used to detect chemicals for medical diagnosis. The sensitivity of a sensor that detects analytes by interacting with molecules on the surface will increase with increasing surface area per unit mass. Therefore, considerable effort has been made to increase the surface area of the sensing interface in chemical sensors [13, 58]. The characteristics possessed by electrospunnanofibers match well with these requirements. Therefore, a nanofibrous structure is a promising physical structure to form a highly sensitive and fast response sensor. Nanofibers with sensing capability have been fabricated by electrospinning polymeric sensing materials, coating/grafting the nanofiber surface with the sensing material, or incorporating the sensing material into nanofibers. For example, pyrene methanol as a sensing material was grafted with polyacrylic and employed for the detection of Fe^{3+} and Hg^{2+} and an explosive 2,4-dinitrotoluene in water [77, 78]. Due to the quenching effect of these chemicals to the pyrene moieties, the fluorescent intensity of nanofiber had a linear response to the concentration of quenchers, and the nanofibers showed high sensitivities [77, 78]. Investigators equally detected methyl viologen and cytochrome C in aqueous solution [77, 78] using a layer-by-layer electrostatically assembled fluorescence optical sensors, whereas trinitrotoluene vapor had been traced by employing a porphyrin-doped silica nanofibers [79]. These nanofibers exhibited fast response as well as high sensitivity. Besides fluorescent properties, conjugated polymer-embedded electrospunnanofibers were also reported to be able to sense volatile organic compounds such as organochlorines based on optical absorption properties [80].

1.2.10. Electrospunnanofibers as ultrafilters

It has been proven that electrospunnanofiber mats were extremely efficient at trapping airborne particles [1, 2, 14] and a very small layer of electrospunnanofiber sprayed onto a porous substrate was sufficient to eliminate the particle penetration. Also, electrospun layers present minimal impedance to moisture vapor diffusion, which is the basis on which they are used as protective clothing in decontamination applications. Electrospunnanofiber membranes are believed to provide significant increase in filtration efficiency at relatively small decreases in permeability. Kosmider and Scott proposed that when compared with conventional filter fibers at the same pressure drop, nanofibers exhibited superior efficiency owing to their ability to collect thinner, finer fibers as the slip flow round the nanofibers enhances the diffusion, interception, and inertia impaction efficiencies [81]. With the aid of 300 nm test particles, thin electrospun nylon-6 membrane (thickness 100 μm, pore size 0.24 μm) also displayed higher filtration efficiency than a commercial highly efficient particulate air (HEPA) filter (thickness 500 μm, pore size 1.7 μm) [82].

1.2.11. Electrospunnanofibers as affinity membranes

Affinity membrane deals with the purification of molecules on the basis of their physical/chemical properties or biological functions instead of their molecular weights/sizes. The principle involves selectively capturing molecules by immobilizing specific ligands onto the membrane surface as opposed to mere sieving. Electrospunnanofibers have a great potential to be functionalized via incorporation of functional materials into the fibers, or via surface chemistry and coating techniques. Interestingly, Affinity membrane combines the exceptional selectivity of the chromatography resins and the reduced pressure drops associated with filtration membranes [10, 11]; thus, it materializes the advancement in both fixed-bed liquid chromatography and membrane filtration. This technology has the merits of lower mass transfer limitation, reduced pressure drops, and enhanced flow rate and productivities when compared to conventional particle-packed column chromatography. When Ma and coworkers surface functionalized electrospun cellulose nanofibers with a dye Cibacron Blue F3GA (CB), the functionalized nanofiber membrane exhibited a strong affinity to bovine serum albumin (BSA) and bilirubin, with a capture ability of 13 mg and 4 mg per gram nanofibers, respectively [83].

1.2.12. Electrospun nanofibrous scaffolds-engineered tissues

The sophisticated architecture of body tissues/organs with their various layers makes it difficult for current synthetic 3D matrices to simulate such processes. Hence, the advent of electrospinning had further simplified the fabrication of both natural and synthetic polymers into biomaterial scaffolds that more closely mimic the tissue matrices. For example, PCL, collagen type I, and collage type I-coated PCL nanofibers had been successfully electrospun to fabricate a substitute for skin regeneration. The results proved that PCL fibers are able to partially support the growth of skin fibroblasts, while the presence of collagen on the scaffolds greatly enhances the interactions between cells and nanofibers [84]. Silk fibroin nanofibers [53], PLGA, and chitin/PLGA nanofibrous scaffolds cultured with keratinocytes and fibroblasts have also shown great promise in skin tissue engineering [8].

Electrospunnanofibers have extensively been investigated in the fabrication of tissue engineered artery. By assembling two electrospun nanofibrous tubes composed of different collagen/elastin ratios with cultured dermal fibroblasts, aortic smooth muscle cells, and umbilical vein endothelial cells in the outer, middle, and inner layers of the scaffold, respectively, Boland et al. fabricated a vascular scaffold that successfully simulated the anatomical three layers of an artery [85]. One major drawback of the study was that the mechanical integrity of the scaffolds was compromised unlike the mammalian artery. In an attempt to compliment for the mechanical integrity, researchers blended PLGA with collagen and elastin to produce an electrospun tripolymeric fibrous scaffold, therefore increasing the strength of the original collagen/elastin scaffold and rendering it mechanically comparable to a native artery [86].

The development of cartilage tissue engineering using the electrospinning technique had been reported [87]. In their study, the biological response of chondrocytes seeded onto 3D PCL nanofibrous scaffolds was compared to that of cells seeded as monolayers on standard tissue culture polystyrene (TCPS) [87]. Gene expression analysis revealed that the chondrocytes exhibited a round shape on the nanofibrous scaffolds, in contrast to a flat, well-spread morphology seen in monolayer cultures on TCPS. They concluded that the biological activities of chondrocytes are significantly dependent on the dimensionality of the extracellular scaffolds and that nanofibrous PCL may be a biologically preferred scaffold/substrate for proliferation and phenotype maintenance of chondrocytes [8, 87].

It is believed that the MSCs for bone tissue engineering must be seeded within electrospun nanofibrous scaffolds in order to promote osteogenic differentiation [8]. On seeding rat bone marrow MSCs in PCL nanofibrous scaffolds and culturing them in a rotatory oxygen-permeable bioreactor with an osteogenic medium, Yoshimoto et al. and Shin et al. demonstrated the osteogenic differentiation of MSCs in PCL nanofibrous scaffolds cultured *in vitro* and *in vivo* [88, 89].

It is worthy to note that a suitable tissue engineered muscle scaffold should be flexible in structure for cardiomyocyte contraction, sustain a reasonable tension for cell morphology maintenance, and have a good integrity for handling. In this line of thought, Shin et al. have successfully cultured primary cardiomyocytes from rat ventricles on PCL nanofiber suspended wire rings [90]. Their results suggested that the dimensionality of the extracellular scaffolds significantly influenced the biological activities of chondrocytes and that the PCL nanofibers served as the better biologically substrate for the cell proliferation and maintenance of the chondrocyte phenotype [90].

An investigation of the effects of fiber alignment on the ECM generation of human ligament fibroblasts (HLF) on polyurethane (PU) electrospunnanofibers revealed that cells cultured on aligned nanofibers were spindle shaped and oriented in the nanofiber direction, whereas cells on nonaligned nanofibers had no directionality [65, 91]. Taken together, the researchers concluded from the results that the biomimetic nature of aligned electrospunnanofibers provides an architectural environment similar to that which ligament fibroblasts normally encounter *in vivo*.

Finally, studies have equally investigated the potential of electrospun 3D scaffolds, both aligned and nonaligned, in neural/nerve tissue engineering employing a multipotent neural stem cell (NSC) line, C17–2, derived from a neonatal mouse cerebellum [8, 65]. Significant changes in the phenotype of cells based on directionality were reported and these results show a significant relationship of a decrease in fiber diameter increasing neurite outgrowth [8, 65]. Since successful nerve regeneration is dependent upon extensive growth of axonal processes, electrospinning presents a sophisticated technique to fabricate an ECM-like nerve tissue.

1.3. Challenges, prospects, and conclusion

It is no gainsaying that the advent of electrospinning has revolutionized the intrigues and applications of nanostructured materials. Several potential applications of electrospunnanofibers have been extensively investigated in separation science, tissue engineering, regenerative engineering, and other clinical fields. Despite the overwhelming benefits of electrospinning, there are still limitations in translating these fabricated electrospun nanomaterials from bench to bedside and/or fabricating human tissues using nanofibrous scaffolds. In the recent past, researchers have introduced a plethora of different new polymers for electrospinning, as well as various characterization techniques; nevertheless, there is limited progress in their biological evaluation. More recent studies are investigating cellular and molecular analyses of cell-cell and cell-nanofiber interactions based on the structural and functional resemblance of electrospunnanofiber scaffolds to native ECM. In addition to actively inducing favorable biological activities, emerging functional electrospun nanofibrous scaffolds are expected to provide structural and mechanical support for tissue regeneration/engineering. Hence, more extensive biological analyses as well as physical characterization are required to determine the biocompatibility of the fabricated nanofiber toward promoting of tissue growth. Owing to the fact that bioactive motifs, peptides, and growth factors are capable of eliciting cellular response, efforts should be channeled toward developing methodologies that can incorporate and optimize bioactive motifs or peptides or growth factors into the electrospun nanofibrous scaffold, thereby creating biologically active scaffolds [8]. Electrospinning is one such strategy that has been widely employed for the successful incorporation of the aforesaid bioactive compounds into nanofibrous scaffolds, as well as controlling the fiber orientation. Owing to the significance of fiber arrangement, electrospinning has a leading edge over other nanofiber fabrication techniques due to the fact that it is able to easily control the orientation of the nanofibers, despite the inherent advantages of conventional nanofiber fabrication techniques [1, 2, 8, 10–14]. It is important to note that the electrospinning process depends on several parameters, and the precise control of each parameter directly affects the morphology of the nanofibers [1, 2, 8, 10–14]. Given the expected complexity of *in vivo* nanofiber scaffolds, obtaining such biocompatibility, biodegradability, nontoxicity, and structural integrity scaffolds precisely using traditional electrospinning technique is challenging due to the unplanned randomly intertwined nanofibers [92]. To precisely control the fiber orientation and electrospinning diameter to produce thinner 3D fibers, an in-depth understanding of controlled fabrication, electrospinning parameters, properties, and functioning of electrospun materials is required to overcome the limitations. The current trend is the emerging robotics technology (3D printing, 3D bioplotting, nanoimprinting, etc.) that has immensely benefited the biofabrication process by improving the flexibility, accuracy, controllability, process parameters, nanofiber diameter, and the rate of nanofibers produced [92]. Currently, there is an overwhelming application of electrospunnanofiber scaffolds/sorbents in analytical field and separations science including cosmetics, as filter media, solid phase extraction (SPE) sorbent bed, purification devices, preconcentration devices, protective clothing, wound dressing, sensor devices, and healthcare systems [1, 2, 10, 11, 58, 93]. Conversely, there have been limited progress in the clinical translation of these nanofibers and the fabricating technique; thus, the future of electrospinning will tend toward drug delivery, cell delivery, gene delivery, DNA/plasmid delivery, tissue engineering, and regenerative engineering against important health challenges. Overall, the role of electrospinning in separation science and biomedical and clinical application cannot be overemphasized. In the near future, the advent of robotics-assisted electrospinning technique will certainly drive nanotechnology from laboratory/industry toward patient care/bedside.

Author details

Okechukwu Clinton Ifegwu[1]* and Chimezie Anyakora[1,2]

*Address all correspondence to: ifegwuclinton@gmail.com

1 The Centre for Applied Research on Separation Science (Caross), Lagos, Nigeria

2 Department of Pharmaceutical Chemistry, Faculty of Pharmacy, University of Lagos, Nigeria

References

[1] Ifegwu OC, Anyakora C, Chigome S, Torto N. Electrospunnanofiber sorbents for the pre-concentration of urinary 1-hydroxypyrene. Journal of Analytical Science and Technology. 2015;**6**(1):1-10

[2] Ifegwu OC, Anyakora C, Chigome S, Torto N. Application of nanofiber-packed SPE for determination of urinary 1-hydroxypyrene level using HPLC. Analytical Chemistry Insights. 2014;**9**:17-25

[3] Huang ZM, Zhang YZ, Kotaki M, Ramakrishna S. A review on polymer nanofibers by electrospinning and their applications in nanocomposites. Composites Science and Technology. 2003;**63**:2223-2253

[4] Utreja P, Jain S, Tiwary AK. Novel drug delivery systems for sustained and targeted delivery of anti-cancer drugs: Current status and future prospects. Current Drug Delivery. 2010;**7**:152-161

[5] Liu Z, Kang X, Fang F. Solid phase extraction with electrospunnanofibers for determination of retinol and alpha-tocopherol in plasma. Microchimica Acta. 2010;**168**:59-64

[6] Kang X-J, Chen L-Q, Wang Y, Zhang Y-Y, Gu Z-Z. Design of packed-fiber solid-phase extraction device for analysis of the drug and its metabolite in plasma. Biomedical Microdevices. 2009;**11**(4):723-729

[7] Kim ID, Choi SJ, Ryu WH. Electrospun functional nanofibers and their applications in chemical sensors and li-ion batteries. In: Bhushan B, Luo D, Schricker S, Sigmund W, Zauscher S, editors. Handbook of Nanomaterials Properties. Berlin, Heidelberg: Springer; 2014

[8] Li W, Shanti RM, Tuan RS. Electrospinning Technology for Nanofibrous Scaffolds in tissue engineering. In: Kumar CSSR, editor. Nanotechnologies for the Life Sciences Vol. 9. Tissue, Cell and Organ Engineering. Copyright 8 2006. Weinheim: WILEY-VCH Verlag GmbH & Co. KGaA; 2006 ISBN: 3-527-31389-3

[9] Bölgen N, Vaseashta A. Nanofibers for tissue engineering and regenerative medicine. In: Sontea V, Tiginyanu I, editors. 3rd International Conference on Nanotechnologies and Biomedical Engineering. IFMBE Proceedings. Vol. 55. Singapore: Springer; 2016

[10] Chigome S, Torto N. A review of opportunities for electrospunnanofibers in analytical chemistry. Analytica Chimica Acta. 2011b;**706**:25-36

[11] Chigome S, Torto N. Electrospun nanofibre-based solid-phased extraction. Trends in Analytical Chemistry. 2012;**2012**:38

[12] Ramakrishna S, Fujihara K, Teo W, Yong T, Ma Z, Ramaseshan R. Electrospunnanofibers: Solving a global issue. Materials Today. 2006;**9**(3)

[13] Teo WE, Ramakrishna S. A review on electro spinning design and nanofibre assemblies. Nanotechnology. 2006;**17**(14):89-104

[14] Ramakrishna S, Fujihara K, Teo W, Lim T, Ma Z. An Introduction to Electrospinning and Nanofibers. Singapore: World Scientific Publishing Co; 2005

[15] Jayaraman K, Kotaki M, Zhang Y, Mo X, Ramakrishna S. Recent advances in polymer nanofibers. Nanoscience and Nanotechnology. 2004;**4**(1-2):52-65

[16] Ondarcuhu T, Joachim C. Drawing a single nanofibre over hundreds of microns. Europhysics Letters. 1998;**42**(2):215-220. DOI: 10.1209/epl/i1998-00233-9

[17] Feng L, Li SH, Li H i, Zhai J, Song YL, Jiang L, Zhu DB. Super hydrophobic surface of aligned polyacrylonitrile nanoflbers. Angewandte Chemie, International Edition. 2002; **41**(7):1221-1223

[18] Yan X, Liu G, Li Z. Preparation and phase segregation of block copolymer nanotube multiblocks. Journal of the American Chemical Society. 2004;**126**(32):10059-10066

[19] Hartgerink JD, Beniash E, Stupp SI. Self-assembly and mineralization of peptide-amphiphile nanofibers. Science. 2001;**294**(5547):1684-1688

[20] Ellison JC, Phatak A, Giles DW, Macosko CW, Bates FS. Melt blown nanofibers: Fiber diameter distributions and onset of fiber breakup. Polymer. 2007;**48**:3306-3316

[21] Chen T, Li L, Huang X. Fiber diameter of polybutylene terephthalate melt-blown nonwovens. Journal of Applied Polymer Science. 2005;**97**:1750-1752

[22] Bresee RR. Influence of processing conditions on melt blown web structure: Part 1 -DCD. INJ Spring. 2004:49-55

[23] Zheng XT, Xu HV, Tan YN. Bioinspired design and engineering of functional nanostructured materials for biomedical applications. Advances in Bioinspired and Biomedical Materials. 2017;**2**:123-152. DOI: 10.1021/bk-2017-1253.ch007

[24] Ma PX, Zhang R. Synthetic nano-scale fibrous extracellular matrix. Journal of Biomedical Materials Research. 1999;**46**(1):60-72

[25] Taylor G. Disintegration of water drops in an electric field. Proceedings of the Royal Society of London. Series A: Mathematical and Physical Sciences. 1964;**280**(1382):383-397

[26] Taylor G. Electrically driven jets. Proceedings of the Royal Society of London. Series A: Mathematical and Physical Sciences. 1969;**313**(1515):453-475

[27] Shenoy SL, Bates WD, Wnek G. Correlations between electrospinnability and physical gelation. Polymer. 2005;**46**(21):8990-9004

[28] Sill TJ, von Recum HA. Electrospinning: Applications in drug delivery and tissue engineering. Biomaterials. 2008;**29**(13):1989-2006. DOI: 10.1016/j.biomaterials.2008.01.011; Epub 2008 Feb 20

[29] Reneker DH, Fong H. Polymeric nanofibers: Introduction. In: Reneker DH, Fong H, editors. Polymeric Nanofibers. ACS Symposium Series 918. USA: Oxford University Press; 2006. p. 430

[30] Deitzel JM, Krauthauser C, Harris D, Perganis C, Kleinmeyer J. Key parameters influencing the onset and maintenance of electro spinning jet. In: Polymeric Nanofibers. ACS Symposium Series 918. USA: Oxford University Press; 2006. p. 56

[31] Lei T, Peng Q, Chen Q, Xiong J, Zhang F, Sun D. Alignment of electrospun fibers using the whipping instability. Materials Letters. 2017;**193**:248-250

[32] Yuan H, Zhou Q, Zhang Y. Improving fiber alignment during electrospinning. Electrospunnanofibers. 2017;**2017**:125-147

[33] Dzenis YA. Spinning continuous fibers for nanoteclmology. Science. 2004;**304**(5679):1917

[34] Wei M, Kang BW, Sung CM, Mead J. Preparation of nanofibers with controlled phase morphology from electrospinning of polybutadiene-polycarbonate blends. In: Reneker DH, Fang H, editors. Polymeric Nanofibers. ACS Symposium Series 918. USA: Oxford University Press; 2006b. p. 149

[35] Larsen G, Spretz R, Velarde-Ortiz R. Use of coaxial gas jackets to stabilize Taylor cones of volatile solutions and to induce particle-to-fiber transitions. Advanced Materials. 2004b;**16**(2):166-169

[36] Kidoaki S, Kwon LK, Matsuda T. Mesoscopic spatial designs of nano- and micro fiber meshes for tissue-engineering matrix and scaffold based on newly devised multilayering and mixing electrospinning techniques. Biomaterials. 2005;**26**(1):37-46

[37] Niu H, Wang X, Lin T. Needleless Electrospinning: Developments and Performances, Nanotechnology and Nanomaterials: Nanofibers - Production, Properties and Functional Applications; 2011

[38] Akduman C, Perrin E, Kumabasar A, Çay A. Effect of molecular weight on the morphology of electrospun poly(vinyl alcohol) nanofibers. XIIIth International Izmir Textile and Apparel Symposium April 2-5; 2014

[39] Megelski S, Stephens JS, Chase DB, Rabolt JF. Micro- and nanostructured surface morphology on electrospun polymer fibers. Macromolecules. 2002;**35**(22):8456-8466

[40] Deitzel JM, Kleinmeyer JD, Hirvonen JK, Tan NCB. Controlled deposition of electrospun poly(ethyl-ene oxide) fibers. Polymer. 2001;**42**:8163

[41] Zhong XH, Kim KS, Fang DF, Ran SF, Hsiao BS, Chu B. Structure and process relationship of electrospun bioabsorbable nanofiber membranes. Polymer. 2002;**43**:4403-4412

[42] Krishnappa RVN, Desai K, Sung C. Morphological study of electrospun polycarbonates as a function of the solvent and processing voltage. Journal of Materials Science. 2003;**38**(11):2357-2365

[43] Zhao SL, Wu XH, Wang LG, Huang Y. Electrospinning of ethyl-Cyanoethyl cellulose/Tetrahydrofuran solutions. Journal of Applied Polymer Science. 2004;**91**:242-246

[44] Patel C, Li S, Wang C, Zhang W, Wei Y. Electrospinning of porous silica nanofibers containing silver nanoparticles for catalytic applications. Chemistry of Materials. 2007;**19**:1231-1238

[45] Mit-uppatham C, Nithitanakul M, Supaphol P. Ultrafine electrospun polyamide-6 fibers: Effect of solution conditions on morphology and average fiber diameter. Macromolecular Chemistry and Physics. 2004b;**205**(17):2327-2338

[46] Khil M, Bhattarai SR, Kim H, Kim S, Lee K. Novel fabricated matrix via electrospinning for tissue engineering. Journal of Biomedical Materials Research Part B. 2005;**72**(1):117-124

[47] Srinivasan G, Reneker DH. Structure and morphology of small diameter electrospun aramid fibers. Polymer International. 1995;**36**(2):195-201

[48] Wannatong L, Sirivat A, Supaphol P. Effects of solvents on electrospun polymeric fibers: Prelinlinary study on polystyrene. Polymer International. 2004;**53**(11):1851-1859

[49] Lee JS, Choi KH, Ghim HD, Kim SS, Chun DH, Kim HY, Lyoo WS. Role of molecular weight of atactic poly(vinyl alcohol) (PVA) in the structure and properties of PVA nanofabric prepared by electro spinning. Journal of Applied Polymer Science. 2004; **93**(4):1638-1646

[50] Yarin AL, Zussman E. Upward needleless electro spinning of multiple nanofibers. Polymer. 2004;**45**(9):2977-2980

[51] Li D, Ouyang G, McCann JT, Xia Y. Collecting electrospunnanofibers with patterned electrodes. Nano Letters ASAP article. 2005;**5**(5):913-916

[52] Son WK, Youk JH, Lee TS, Park WH. Electrospinning of ultrafme cellulose acetate fibers: Studies of a new solvent system and deacetylation of ultrafme cellulose acetate fibers. Journal of Polymer Science Part B: Polymer Physics. 2004;**42**:5-11

[53] Min BM, Lee G, Kim SH, Nam YS, Lee TS, Park WH. Electrospinning of silk fibroin nanofibers and its effect on the adhesion and spreading of normal human keratinocytes and fibroblasts in vitro. Biomaterials. 2004;**25**:1289-1297

[54] Casper CL, Stephens JS, Tassi NG, Chase DB, Rabolt JF. Controlling surface morphology of electrospun polystyrene fibers: Effect of humidity and molecular weight in the electrospinning process. Macromolecules. 2004;**37**:573-578

[55] Fasano V, Moffa M, Camposeo A, Persano L, Pisignano D. Controlled atmosphere electrospinning of organic nanofibers with improved light emission and waveguiding properties. Macromolecules. 2015;**48**:7803-7809

[56] Um IC, Fang DF, Hsiao BS, Okamoto A, Chu B. Electro-spinning and electro-blowing of hyaluronic acid. Biomacromolecules. 2004;**5**:1428-1436

[57] Subramanian A, Vu D, Larsen GF, Lin HY. Preparation and evaluation of the electrospun chitosan JPEO fibers for potential applications in cartilage tissue engineering. Journal of Biomaterials Science. 2005;**16**(7):861-873

[58] Ifegwu OC, Anyakora C, Torto N. Nylon 6-gold nanoparticle composite fibers for colorimetric detection of urinary 1-hydroxypyrene. Journal of Applied Spectroscopy. 2015;**82**(2):260-265

[59] Kim K, Luuc YK, Chang C, Fang D, Hsiao BS, Chua B, Hadjiargyrou M. Incorporation and controlled release of a hydrophilic antibiotic using poly(lactide-co-glycolide) based electrospun nanofibrous scaffolds. Journal of Controlled Release. 2004;**98**:47-56

[60] Doshi J, Reneker DH. Electrospinning process and application of electro spun fibers. Journal of Electrostatics. 1995;**35**:151-160

[61] Nikolovski J, Mooney DJ. Smooth muscle cell adhesion to tissue engineering scaffolds. Biomaterials. 2000;**21**:2025-2032

[62] Woo KM, Chen VJ, Ma PX. Nanofibrous scaffolding architecture selectively enhances protein adsorption contributing to cell attachment. Journal of Biomedical Materials Research. 2003;**67A**:531-537

[63] Scully SP, Lee JW, Ghert PMA, Qi W. The role of the extracellular matrix in articular chondrocyte regulation. Clinical Orthopaedics. 2001:S72-S89

[64] Li WJ, Tuli R, Huang X, Laquerriere P, Tuan RS. Multilineage differentiation of human mesenchymal stem cells in a three-dimensional nanofibrous scaffold. Biomaterials. 2005;**26**:5158-5166

[65] Yang F, Xu CY, Kotaki M, Wang S, Ramakrishna S. Characterization of neural stem cells on electrospun poly(L-lactic acid) nanofibrous scaffold. Journal of Biomaterials Science, Polymer Edition. 2004;**15**(12):1483-1497

[66] Li S, Edgar D, Fassler R, Wadsworth W, Yurchenco PD. The role of laminin in embryonic cell polarization and tissue organization. Developmental Cell. 2003;**4**:613-624

[67] Nur-E-Kamal A, Ahmed I, Kamal J, Schindler M, Meiners S. Threedimensional nanofibrillar surfaces promote self-renewal in mouse embryonic stem cells. Stem Cells. 2005;**24**:426-433

[68] Yang F, Murugan R, Wang S, Ramakrishna S. Electrospinning of nano/micro scale poly(L-lactic acid) aligned fibers and their potential in neural tissue engineering. Biomaterials. 2005;**26**:2603-2610

[69] Garg T, Rath G, Goyal AK. Biomaterials-based nanofiber scaffold: Targeted and controlled carrier for cell and drug delivery. Journal of Drug Targeting. 2014:1-20. Informa UK Ltd. DOI: 10.3109/1061186X.2014.992899

[70] Jia H, Zhu G, Vugrinovich B, Kataphinan W, Reneker DH, Wang P. Enzyme-carrying polymeric nanofibers prepared via electrospinning for use as unique biocatalysts. Biotechnology Progress. 2002;**18**:1027-1032

[71] Huang CB, Chen SL, Reneker DH, Lai CL, Hou H. High strength mats from electrospun poly(p-phenylene biphenyltetracarboximide) nanofibers. Advanced Materials. 2006b;**18**(5):668-671

[72] Huang CB, Chen SL, Lai CL, Reneker DH, Qiu HY, Ye Y, Hou HQ. Electrospun polymer nanofibres with small diameters. Nanotechnology. 2006a;**17**(6):1558-1563

[73] Herricks TE, Kim S-H, Kim J, Li D, Kwak JH, Grate JW, Kim SH, Xia Y. Direct fabrication of enzyme-carrying polymer nanofibers by electrospinning. Journal of Materials Chemistry. 2005;**15**(31):3241-3245

[74] Yu J, Liu T. Preparation of nano-fiber supported palladium catalysts and their use for the catalytic hydrogenation of olefins. Acta Polymerica Sinica. 2007;**6**:514-518

[75] Choi SK, Kim S, Lim SK, Park H. Photocatalytic comparison of TiO_2 nanoparticles and electrospun TiO_2 nanofibers: Effects of mesoporosity and interparticle charge transfer. Journal of Physical Chemistry C. 2010c;**114**(39):16475-16480

[76] Zhang Z, Shao C, Zhang L, Li X, Liu Y. Electrospunnanofibers of V-doped TiO_2 with high photocatalytic activity. Journal of Colloid and Interface Science. 2010c;**351**(1):57-62

[77] Wang CE, Zhen YL, Li D, Yang QB, Hong YL. Preparation and stability of the nanochains consisting of copper nanoparticles and PVA nanofiber. International Journal of Nanoscience. 2002a;**1**(5-6):471-476

[78] Wang X, Drew C, Lee SH, Senecal KJ, Kumar J, Samuelson LA. Electrospun nanofibrous membranes for highly sensitive optical sensors. Nano Letters. 2002b;**2**(1):1273-1275

[79] Tao S, Li G, Yin J. Fluorescent nanofibrous membranes for trace detection of TNT vapor. Journal of Materials Chemistry. 2007;**17**(26):2730-2736

[80] Yoon J, Chae SK, Kim JM. Colorimetric sensors for volatile organic compounds (VOCs) based on conjugated polymer-embedded electrospun fibers. Journal of the American Chemical Society. 2007;**129**(11):3038-3039

[81] Kosmider K, Scott J. Polymeric nanofibres exhibit an enhanced air filtration performance. Filtration & Separation. 2002;**39**(6):20-22

[82] Barhate RS, Ramakrishna S. Nanofibrous filtering media: Filtration problems and solutions from tiny materials. Journal of Membrane Science. 2007;**296**(1-2):1-8

[83] Ma Z, Kotaki M, Ramakrishna S. Surface modified nonwoven poly-Sulphone (PSU) fiber mesh by electrospinning: A novel affinity membrane. Journal of Membrane Science. 2006; **272**(1-2):179-187

[84] Venugopal J, Ramakrishna S. Biocompatible nanofiber matrices for the engineering of a dermal substitute for skin regeneration. Tissue Engineering. 2005;**11**:847-854

[85] Boland ED, Matthews JA, Pawlowski KJ, Simpson DG, Wnek GE, Bowlin GL. Electrospinning collagen and elastin: Preliminary vascular tissue engineering. Frontiers in Bioscience. 2004;**9**:1422-1432

[86] Stitzel J, Liu J, Lee SJ, Komura M, Berry J, Soker S, Lim G, Van Dyke M, Czerw R, Yoo JJ, Atala A. Controlled fabrication of a biological vascular substitute. Biomaterials. 2006;**27**:1088-1094

[87] Li WJ, Danielson KG, Alexander PG, Tuan RS. Biological response of chondrocytes cultured in threedimensional nanofibrous poly(epsiloncaprolactone) scaffolds. Journal of Biomedical Materials Research. Part A. 2003;**67**:1105-1114

[88] Yoshimoto H, Shin YM, Terai H, Vacanti JP. A biodegradable nanofiber scaffold by electrospinning and its potential for bone tissue engineering. Biomaterials. 2003;**24**:2077-2082

[89] Shin M, Yoshimoto H, Vacanti JP. In vivo bone tissue engineering using mesenchymal stem cells on a novel electrospun nanofibrous scaffold. Tissue Engineering. 2004;**10**:33-41

[90] Shin M, Ishii O, Sueda T, Vacanti JP. Contractile cardiac grafts using a novel nanofibrous mesh. Biomaterials. 2004;**25**:3717-3723

[91] Lee CH, Shin HJ, Cho IH, Kang YM, Kim IA, Park KD, Shin JW. Nanofiber alignment and direction of mechanical strain affect the ECM production of human ACL fibroblast. Biomaterials. 2005;**26**:1261-1270

[92] Tan R, Yang X, Shen Y. Robot-aided electrospinning toward intelligent biomedical engineering. Robotics and Biomimetics. 2017;**4**:17. DOI: 10.1186/s40638-017-0075-1

[93] Fang J, Wang H, Niu H, Lin T, Wang X. Evolution of fiber morphology during electrospinning. Journal of Applied Polymer Science. 2010;**118**(5):2553-2561

Chapter 4

Electrospun Poly(Ethylene Oxide) Fibers Reinforced with Poly (Vinylpyrrolidone) Polymer and Cellulose Nanocrystals

Qinglin Wu, Changtong Mei, Xiuqiang Zhang,
Tingzhou Lei, Zhen Zhang and Meichun Li

Additional information is available at the end of the chapter

http://dx.doi.org/10.5772/intechopen.76392

Abstract

Green poly(ethylene oxide) (PEO)/cellulose nanocrystals (CNCs)/poly(vinylpyrrolidone) (PVP) composites were prepared via electrospinning technique. The use of PVP and/or CNCs improved the overall thermal stability and mechanical properties of the PEO fibers. A strong synergistic reinforcing effect was achieved when PVP polymer and CNCs were combined in the composite. This synergistic reinforcement was accompanied with the formation of unique fiber-bead-fiber morphology. The beads were elongated and orientated along the applied force direction during tensile testing, providing an energy dissipation mechanism and a positive reinforcement effect. The combination of CNCs with PVP induced special chemical interactions, and distracted the interactions between PVP and PEO. As a result, the crystallinity of PEO was increased in the system, which also helped enhance fiber properties. The approach developed in this work offers a new way for reinforcing electrospun PEO-based composite fibers for sustainable green composite development.

Keywords: electrospinning, cellulose nanocrystals, reinforcing mechanisms, poly(ethylene oxide), poly(vinylpyrrolidone), green composites

1. Introduction

Currently, fabricating "green polymeric materials" that are environmentally friendly has received much attention in academic and industrial research [1–3]. However, one drawback is relatively low strength of these biopolymers, which restricts their wide use for replacing synthetic polymers [4, 5]. One effective way to improve their strength and commercial use potential is to incorporate nano-sized reinforcement into the polymeric matrix [6–8]. Among the nano-fillers, cellulose nanocrystals (CNCs) have been considered as a competitive candidate, because of their high strength and Young's modulus [6, 9]. In addition, CNCs are inherently renewable, biodegradable, and biocompatible, leading to long-term sustainability [10, 11].

Poly(ethylene oxide) (PEO), as a nontoxic, biodegradable, and biocompatible polymer, has received tremendous attention in biomedical field [12, 13]. PEO-water solution system can be used to prepare nanofibers via electrospinning technique [6, 14, 15] for bone regeneration [16], drug release delivery [17], and adsorption of ions [18]. However, low mechanical strength of electrospun PEO nanofibers generally limits their application. To reinforce PEO nanofibers without decreasing their biodegradability and biocompatibility, green materials such as CNCs and cellulose nanofibers (CNFs) have been recently investigated. Zhou et al. [6] found that in the PEO/CNC binary nanocomposite system, CNCs can be highly aligned under high electrostatic fields, and induced heterogeneous microstructures regarding the uniformity of fiber size, which resulted in a considerable increase in mechanical strength. In another study [14], both CNCs and cellulose nanofibers (CNFs) were used to reinforce electrospun PEO nanofibers and CNFs exhibited a higher reinforcing efficiency, which was attributed to the unique shish-kebab-like crystalline structures formed within the nanofibers. The reinforcing effect of CNCs for PEO composite fibers needs to be further elucidated to understand their specific role in the composite.

Similar as PEO and CNCs, poly(vinylpyrrolidone) (PVP) is also biodegradable and biocompatible, and has been widely used in biomedical areas [19]. The introduction of PVP to the PEO-CNC system can offer some unique properties without deteriorating the green nature of PEO/CNC system. PVP is a rigid polymer with a glass transition temperature of 160°C [20]. Its higher rigidity in comparison with that of PEO makes it possible to serve as a reinforcement additive. PVP is also an amorphous polymer, which can influence the crystallization of PEO. For semi-crystalline polymers like PEO, its mechanical properties strongly depend upon its crystalline characteristics [21]. Therefore, it is possible to tailor mechanical properties of the material by using PVP in the composite. The carbonyl group in the molecular structure of PVP can work as the proton acceptor, which is prone to chemically interact with the proton in hydroxyl group (–OH) [22]. Considering the –OH groups in the molecular structure of PEO and CNCs, the addition of PVP can possibly contribute to special chemical interactions, which is also beneficial to the reinforcement [23].

Thus, it is of significant interest to incorporate PVP to the PEO-CNC composite via electrospinning technique for tailoring the mechanical properties of the composite system. Herein, electrospun PEO nanofiber mats reinforced with different contents of PVP and CNCs were prepared and a synergistic reinforcing effect was demonstrated in the ternary composites. Rheological properties of the electrospinning materials, thermal stability, crystallization, mechanical properties and morphology of the electrospun mats were investigated. An in-depth study on the reinforcing mechanism was presented.

2. Experimental

2.1. Materials

PEO powder (M_v = 900,000 g/mol) was purchased from Sigma-Aldrich [St. Louis, MO]. PVP (K-30) was provided by Xilong Chemical Reagent Co. Ltd (Guangdong, China). The CNCs were extracted from the cellulose powder provided by Nippon Paper Chemicals Co., LTD (W-50 grade of KC Flock, Tokyo, Japan). In brief, acid hydrolysis of the wood powder with 64 wt% H_2SO_4 followed by high-pressure homogenization was used as described in our previous paper [10]. The specific dimensions of CNCs were 149 in length and 9 nm in diameter, respectively [10]. The prepared CNC suspensions were freeze-dried to get the dry CNC materials.

2.2. Electrospinning

To prepare the PEO/PVP/CNCs suspensions for electrospinning, predetermined amount of PEO, PVP and freeze-dried CNCs were mixed with distilled water, and the mixture was magnetically stirred for several hours, followed by ultrasonic treatment. The total polymeric concentration (i.e., PEO/PVP) was kept constant at 3 wt% for all solutions and suspensions. The weight percentages of PEO, PVP and CNCs were varied for different composite systems (**Table 1**). The prepared samples were designated as PxCy, where x and y represent the dosages of PEO/PVP (polymer) and CNCs, respectively, in part-per-hundred (phr). The prepared materials were loaded in a 5-ml BD plastic syringe with a 25 gauge stainless steel needle. A Chemyx Fusion 100 syringe pump (Stafford, TX) was used to feed the material with controlled feeding rates. A high voltage (15 KV) was applied between the needle and grounded aluminum fiber collector. The fiber collector was covered with a piece of aluminum foil pretreated with a small amount of silicon oil for better fiber mat release. After the electrospinning process, the mats with a thickness around 0.1 mm were removed from the collector plate and vacuum-dried at 40°C for 6 h before characterization.

Composite material type	Polymer system[a]		Reinforcing agent[b]
	PEO (PHR)	PVP (PHR)	CNC (PHR)
P100	100	0	0
P100C5	100	0	5
P90	90	10	0
P90C5	90	10	5
P80	80	20	0
P80C5	80	20	5

[a]PHR: part-per-hundred; Total polymer system (PEO/PVP) made up 100 PHR.
[b]CNC was in relation to the total polymer in the system in PHR.

Table 1. Formulation information of different electrospun PEO-PVP-CNC composite fiber mat systems.

2.3. Characterizations

Rheological properties of the electrospinning solutions/suspensions were investigated using an AR2000ex Rheometer (TA Instruments, New Castle, DE). A 40 mm cone-plate geometry with a cone angle of 1°59′42″ and 52 μm truncation was used for both flow and oscillatory measurements. The testing temperature was kept at 25°C using a temperature-regulated Peltier plate device. In order to avoid water evaporation from test samples, a solvent trap was used to seal the gap between the top cone-plate geometry and bottom Peltier plate. The moat on the top of the trap was filled with silicon oil. For flow measurement, steady shear rates applied varied within 1–100 s^{-1}. With regard to the oscillatory measurement, the angular frequency range was 0.05–100 s^{-1} and the strain was selected to be 50%, which was within the linear viscoelastic region. Fourier transform infrared (FTIR) spectra of the electrospun mats were tested using a Bruker FTIR analyzer (Tensor-27, Bruker Optics, Billerica, MA) with an attenuated total reflectance (ATR) mode. Thirty two scans combined with a spectral range of 4000–600 cm^{-1} were used. Differential scanning calorimetry (DSC) was performed with a TA Q200 DSC system (TA Instruments-Waters LLC, New Castle, DE). Each sample of about 2–6 mg was tested in a hermetic aluminum pan under 40 ml/min nitrogen flow. In order to eliminate any possible thermal history, the samples were first heated to 180°C at a heating rate of 40°C/min. Then, the samples were quickly quenched to −90°C to freeze polymeric chains in the material. The samples were then heated to 180°C at 10°C/min, equilibrated at 180°C for 1 min and cooled down to −40°C at 10°C/min. Thermogravimetric analysis (TGA) was carried out using a Q50 TG analyzer (TA Instruments-Waters LLC, New Castle, DE). All the tests were carried out under nitrogen flow (40 ml/min) in the temperature range of 50–600°C at 10°C/min. X-ray diffraction (XRD) measurements were carried out using a PANalytical empyrean X-ray diffractometer (PANalytical Co., Longmont, CO) with a Cu Kα X-ray resource. All the measurements were operated at 45 kV and 40 mA. Samples were scanned from 4 to 90° and the step size was 0.0263°. Surface morphologies of the electro-spun mats were observed by a scanning electron microscope (SEM, Hitachi High Technologies America, Schaumburg, IL). The surface of each sample was sputter-coated with platinum before viewing. The accelerating voltage was selected to be 20 kV. Tensile properties of the electrospun fiber mats were investigated using an AR2000ex rheometer (TA Instruments-Waters LLC, New Castle, DE) with a solid clamp fixture. Three tensile specimens (20 in length and 5 mm in width) were cut from the electrospun mats. The gauge length was 12 mm, and the testing speed was 1.2 mm/min. Mechanical properties, including tensile strength and elongation at break, were calculated based on the measured stress-strain curves.

3. Results and discussions

3.1. Rheological properties of electrospinning solutions

It is well known that the viscosity of electrospinning solutions plays a key role in the morphology and performance of electrospun mats. **Figure 1a** shows the relationship between the viscosity and shear rate for different materials formulated. For pure PEO solution (P100),

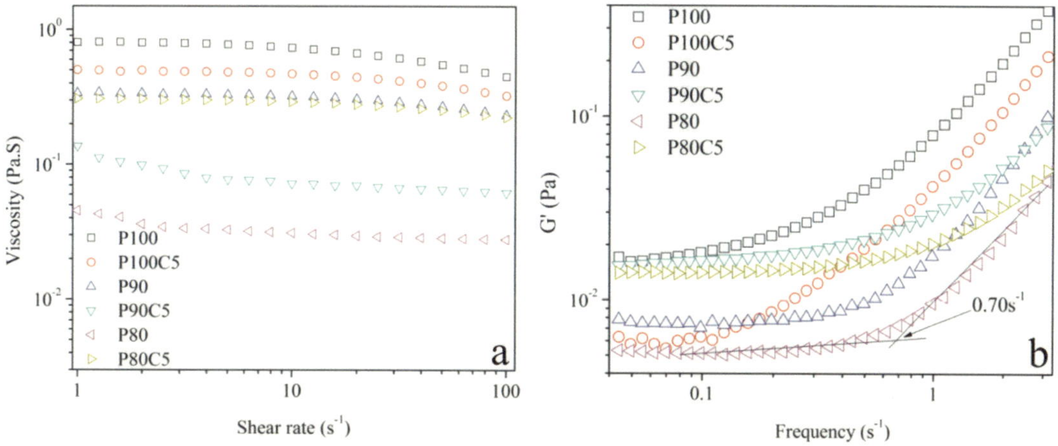

Figure 1. Rheological properties of different suspensions: (a) viscosity-shear rate plot (b) storage modulus-frequency plot from oscillatory measurements.

its viscosity decreased with the increased shear rate, which was attributed to the shear thinning phenomenon [24]. With the addition of PVP and CNCs, the viscosity of the suspensions decreased. For PEO/PVP material, the decrease was attributed to the more viscous nature of PVP. With regard to suspensions that contain CNCs, the decrease was attributed to the interaction between CNCs and PEO. With addition of CNCs, the intra-molecular interaction between PEO molecules was distracted, resulting in the decrease in viscosity. It should be pointed that the viscosity of P80C5 was higher than that of P90C5 and P80, attributed to the special chemical interaction (discussed in the next section).

For oscillatory rheological measurements, the relationship between storage modulus and frequency in the low frequency or terminal region allowed us to evaluate the intermolecular interactions in a multi-component system. If there are certain special chemical interactions (e.g., H-bond interactions) in the tested samples, a so-called "second platform" usually exists [23]. As shown in **Figure 1b**, an obvious platform for P80C5 was observed, which is in a sharp contrast with that for P100 and indicates possible special interactions. To quantitatively compare the length of platform, the extrapolated critical frequency that denotes the frequency at which the substantial increase in storage modulus begins was calculated (e.g., 0.70 s^{-1} for P80 in **Figure 1b**).

P100 showed a critical frequency of 0.29 s^{-1}. With the addition of 5 phr CNCs (P100C5), the critical frequency decreased to 0.11 s^{-1}. Such a decrease was expected, since the hydroxyl groups in CNCs can interact with the ether oxygen in PEO molecules (**Figure 2b**), and therefore distract the intramolecular interactions of PEO (**Figure 2a**) [6]. With the addition of PVP, the critical frequency increased to 0.51 and 0.70 for P90 and P80, respectively, which was attributed to the interaction between carbonyl groups in PVP and hydroxyl groups in PEO (**Figure 2c**). When PVP and CNCs were combined, it is interesting to note that the critical frequency still increased, even considering the fact that CNCs can distract the intramolecular interactions of PEO. This interesting phenomenon was due to the interactions between carbonyl group and hydroxyl groups in CNCs (**Figure 2d**) [22].

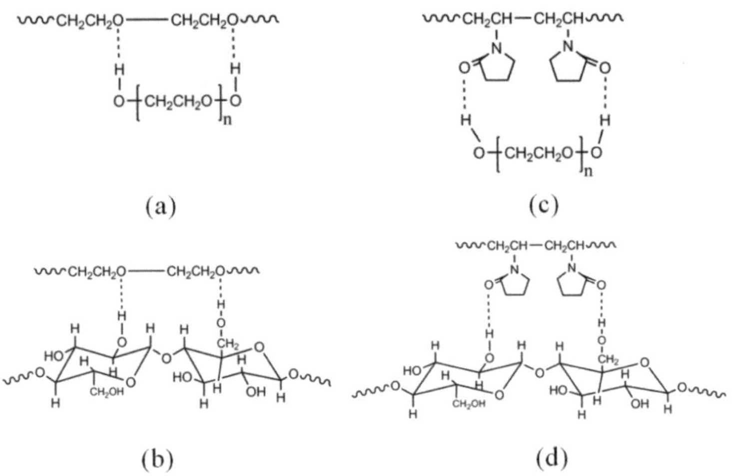

Figure 2. Scheme of interactions between different materials in the composite fibers: (a) PEO/PEO; (b) PEO/PVP; (c) PEO/CNCs; (d) PVP/CNCs.

3.2. FTIR spectra of electrospun mats

FTIR was further used to investigate the proposed interaction and the specific curves are shown in **Figure 3a**. Pure PEO electrospun mat exhibited the absorption bands at 2882, 1467, 1146, 1098, 1060 and 961 cm^{-1}, which were assigned to CH_2 stretching, C–H bending mode and C–O–C stretching vibration, respectively [6]. It should be pointed out that the O–H absorption band was undetectable in FTIR curves. This was due to the high molecular weight of PEO, and therefore the role of its end group (–OH) could be ignored [25]. With the addition of 5 phr CNCs, the absorption band for C–O–C stretching vibration was shifted from 961 to 957 cm^{-1}, suggesting the possible interactions between PEO and CNCs. With respect to PEO/PVP and PEO/PVP/CNCs systems, the main absorption bands were not shifted, but instead new absorption bands assigned to the carbonyl groups in PVP [22] were observed in **Figure 3b**. For P90 and P90C5, the wave number of carbonyl group was located at 1673 cm^{-1}. As the PVP content increased to 20 phr, the peak was shifted to 1669 cm^{-1} for P80 and 1665 cm^{-1} for P80C5, indicating possible interactions existed in these systems [25]. All the FTIR results corresponded well with the abovementioned results of oscillatory rheological tests.

3.3. DSC data of electrospun mats

The effects of CNCs and PVP on the melting and crystallization behaviors of PEO were investigated using the DSC technique. The specific parameters derived from DSC curves, including peak melting, and peak crystallization temperature, melting enthalpy, the half time of crystallization, and crystallinity, are listed in **Table 2**. The crystallinity (X_c) is calculated using the following equation:

$$X_C = \frac{\Delta H_m / \varphi}{\Delta H_m^*} x\,100\% \quad (1)$$

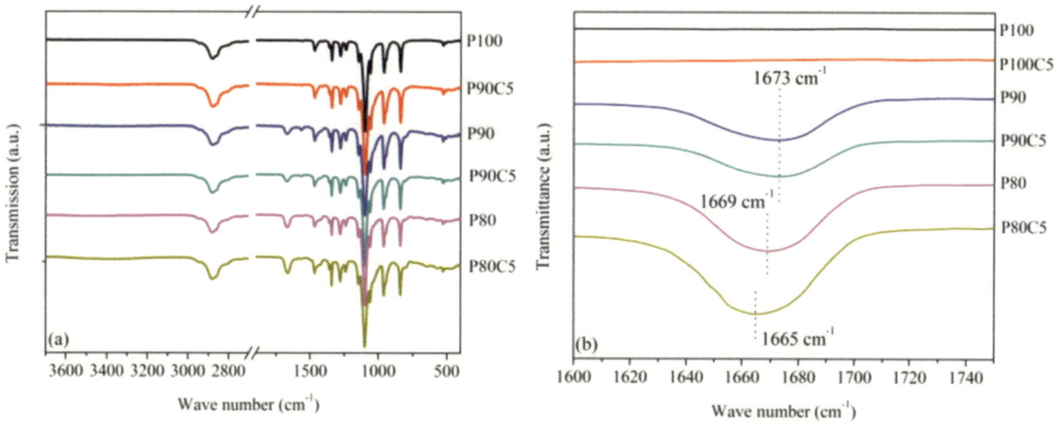

Figure 3. FTIR curves of different electrospun fiber mats: (a) 3700–400 cm^{-1}; (b) 1750–1600 cm^{-1}.

Material type	DSC[a]					TGA[b]			
	T_m^p (°C)	ΔH_m (J·g^{-1})	X_c (%)	T_c^p (°C)	$t_{1/2}$ (min)	T_{onset} (°C)	T_{p1} (°C)	T_{p2} (°C)	Deriv. weight (%/°C)
P100	65.9	157.4	73.7	48.3	0.26	393.5	-	416.6	3.11
P100C5	64.9	141.3	69.5	48.6	0.26	386.9	-	417.2	2.29
P90	64.4	131.4	68.3	47.7	0.27	374.7	-	413.5	1.80
P90C5	63.7	125.3	68.4	49.3	0.20	382.2	278.3	416.5	2.75
P80	64.8	119.6	69.9	48.1	0.26	372.6	288.4	416.7	2.13
P80C5	64.7	120.1	73.9	49.4	0.19	379.9	277.7	419.1	1.83

[a]T_m^p: peak melting temperature; T_c^p: peak crystallization temperature; $t_{1/2}$: the half time of crystallization; ΔH_m: melting enthalpy of PEO; X_c: crystallinity of PEO.

[b]T_{onset} is the temperature corresponding to 10 wt% weight loss; T_{p1} and T_{p2} are the temperature corresponding to the first and the second peak of DTG curve; Deriv. Weight: derive. weight of the second peak of DTG curve.

Table 2. DSC and TGA data of different electrospun PEO-PVP-CNC composite fiber mats systems.

Where ΔH_m is the melting enthalpy derived from the DSC curves, $\Delta H_m^* = 213.7$ J/g [14] is the melting enthalpy when crystallinity of PEO is 100%, and φ is the weight fraction of PEO in the blend and composites.

As shown in **Table 2**, addition of either PVP or CNC decreased the T_m^p and the ΔH_m of PEO. These decreases can be accounted by the possible interactions between different components with PEO. And the slight decrease in T_m^p indicated that the crystallite size of PEO became smaller [26]. With regard to crystallinity, mere incorporation of PVP and CNCs into PEO did cause decrease in X_c. For example, the X_c for P100 decreased from 73.7 to 69.5% and 69.9 for P100C5 and P80, respectively. However, it is interesting to note that similar decrease in X_c did not occur when CNCs and PVP were combined. The X_c of P80C5 was 73.9%, which was even higher than that of P100. This result might be in conflict with the possible interactions observed in the abovementioned oscillatory rheological measurements and FTIR results, since special interactions usually hinders the crystallization and results in a decreasing crystallinity [26]. Similar interesting results were also observed regarding the crystallization behaviors. As shown in **Table 2**, either PVP or CNCs did not change $t_{1/2}$. However, the T_c^p values of P90C5 and P80C5 were shifted to higher temperature while their $t_{1/2}$ value were reduced, indicating that the nucleating effect occurred [27]. These unexpected results in ternary composite system, including the increased crystallinity, and the nucleating effect, were responsible for the improved mechanical performance.

3.4. Thermal stability data of electrospun mats

The effects of PVP and CNCs on the thermal stability of the composite fibers were investigated by TG analysis. The relevant parameters derived are summarized in **Table 2**. Pure PEO nanofibers exhibited one-step degradation with T_{onset} of 393.5°C. With the addition of PVP, the blend (P90) still exhibited one-degradation step, but the T_{onset} decreased considerably. This is due to the relative lower thermal stability of PVP. When CNCs were added, two-step degradation was observed for all the prepared composites, which were assigned to the degradation of CNCs and polymer/blends, respectively. Incorporation of CNCs to the P100 system decreased the T_{onset} values (P100C5). Surprisingly, the addition of CNCs slightly increased the T_{onset} values of P90C5 and P80C5 in comparison with these for P90 and P80, indicating that CNCs even played a positive role in enhancing the initial thermal stability of the P90 and P80 systems. This improvement might be due to the enhanced crystallinity of P90C5 and P80C5 [9]. The peak height of the second peak on the DTG curves decreased with the addition of either PVP or CNCs. This was due to the dilution effect [28, 29] and indicated that the addition of other components could enhance the overall thermal stability of PEO, even though they did decrease the initial thermal stability (T_{onset}).

3.5. Crystalline structure of electrospun mats

XRD was used to investigate the crystal structures of the nanofibers. The diffraction patterns are shown in **Figure 4**. The crystallite size was calculated using the Scherrer equation as follows [6]:

$$L = \frac{K\lambda}{B} \cos\theta \tag{2}$$

where λ is 1.54 Å (the wavelength of Cu K α X-ray), K is 0.89 and B is the full width at half maximum of diffraction peaks. Pure PEO nanofibers exhibited two diffraction peaks at 19° and 23°, which were assigned to the diffraction of the (120, 112) crystal plane [14], respectively. With the addition of PVP and CNC, no shift in the diffraction peak was observed for P80 and P80C5, indicating that the crystal type did not change. However, the crystallite size of (120) plane decreased from 371 Å for P100 to 337 Å for P80C5. This decrease in crystallite size corresponded well with the abovementioned decreasing T_m^p observed in DSC tests.

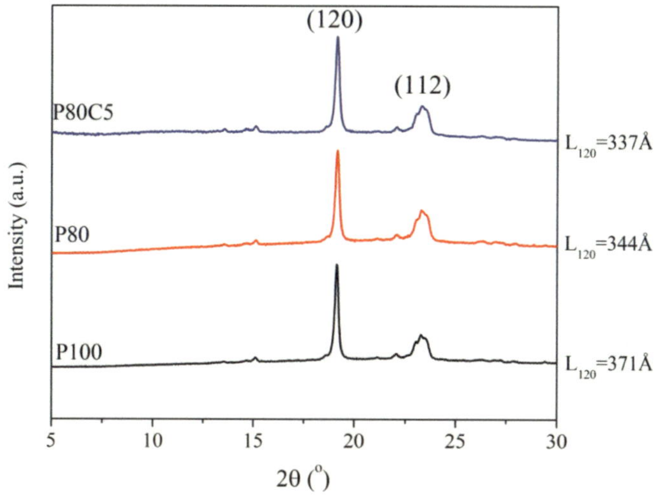

Figure 4. XRD results of different electrospun fiber mats.

3.6. Tensile properties of electrospun mats

The tensile properties, including tensile strength, elongation at break and fracture toughness, are summarized in **Figure 5**. As shown in **Figure 5a**, addition of either PVP or CNCs did have a limited reinforcing effect on PEO. For example, the tensile strength increased from 1.59 MPa (P100) to 1.94 MPa (P100C5) and 2.26 MPa (P80), respectively. Surprisingly, when 20 phr PVP and 5 phr CNCs were combined, an impressive reinforcing effect was achieved. The tensile strength increased to 3.17 MPa, which was nearly a one-fold increase. For elongation at break (**Figure 5b**), addition of PVP and CNCs resulted in a slight decrease. This decline was expected, since the increased rigidity was often accompanied with loss in ductility [9]. With regard to the fracture toughness (**Figure 5b**), all samples exhibited a slight decrease except P80C5. The fracture toughness of P80C5 increased from 223 to 320 kJ/m^2. This was due to the higher strength value of P80C5 (the fracture toughness is proportional with the area under the stress-strain curves) and indicated that the combination of 20 phr PVP and 5 phr CNCs could simultaneously reinforce and toughen PEO.

3.7. Morphology of electrospun mats

The morphologies of different electro-spun mats are shown in **Figure 6**. For pure PEO (P100), the nanofibers combined with some beads were observed, which is different from the homogeneous nanofibers reported in other literatures [6, 14]. This difference is believed to be due to the lower concentrations of PEO used in this work, based on which the entanglement between different macromolecules was low and therefore resulted in some fiber-bead-fiber morphology after electrospinning process [30]. With the addition of CNCs and PVP, more beads were introduced. For the composite systems, addition of CNCs distracted the interactions between PEO molecules, which results in the lower entanglement between PEO molecules. With

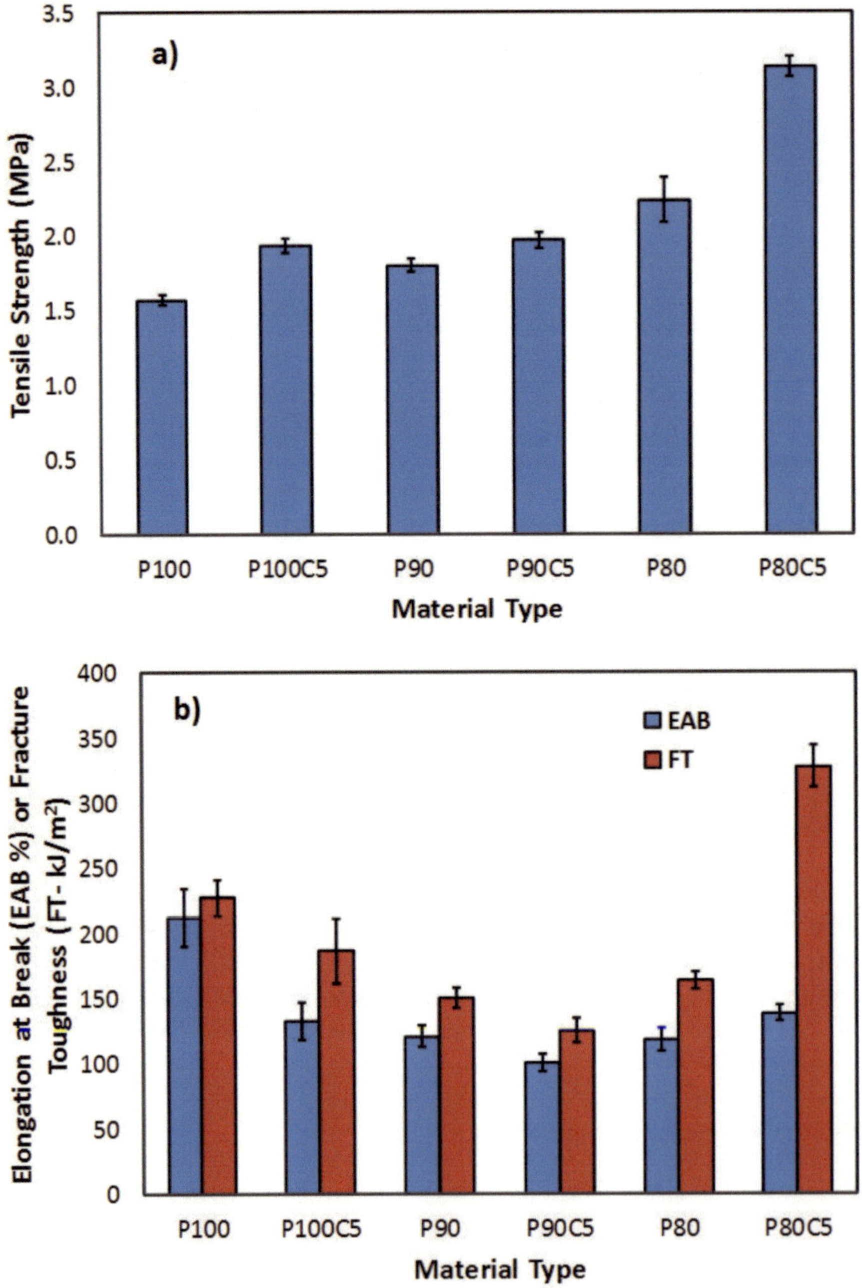

Figure 5. Mechanical properties of different electrospun fiber mats: (a) tensile strength and (b) elongation at break and fracture toughness.

regard to systems that contain PVP, its low molecular weight and low electrospinning nature accounted for the formation of the bead morphology [31]. The electrospun mats with fiber-bead-fiber morphology exhibited the highest mechanical performance as discussed earlier.

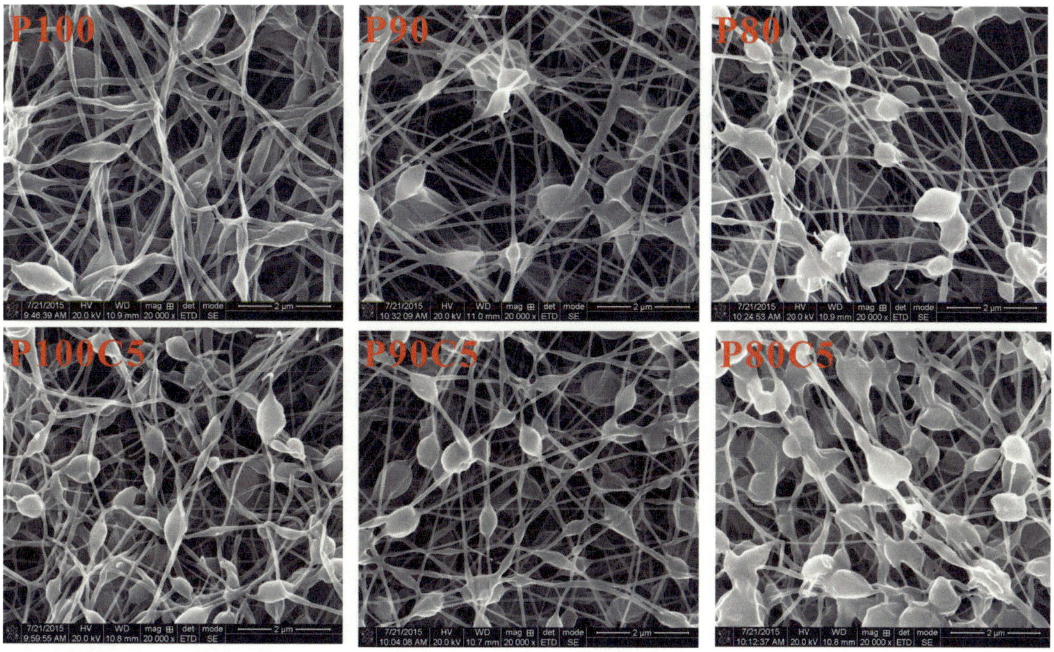

Figure 6. SEM images of different electro-spun fiber mats.

To better elucidate the role of beads played in the fiber reinforcement, the morphology of P80C5 before and after tensile test was obtained (**Figure 7**). The nanofibers were highly orientated along the tensile stress direction (the horizontal direction). It is interesting to note that the number of the beads became smaller and the shape of the beads was highly elongated under stress. This indicated that the beads worked as stress concentrators, and they were elongated, helped dissipate the applied energy, and therefore resulted in overall composite strength reinforcement.

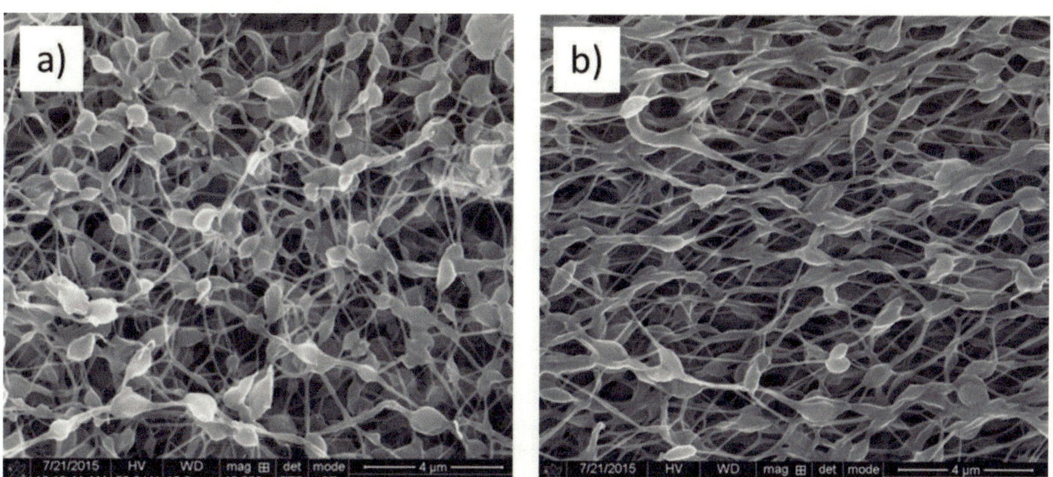

Figure 7. SEM images of the P80C5 composite fiber mat before and after tensile test.

3.8. Discussion on reinforcing mechanisms

There are many factors that determine whether a reinforcing effect can be achieved in the composite fibers. Based on the abovementioned results, the reinforcing mechanism is proposed and its schematic presentation is shown in **Figure 8**. There are three possible factors that contribute to the reinforcement.

The first factor is the crystallinity. For semi-crystalline polymers like PEO, its mechanical strength is proportional with its crystallinity. Herein, in comparison with that of 69.9% for P80, the crystallinity of P80C5 was 73.9%, which was almost the same as that for P100 (73.7%). The reason for the increase in crystallinity is the interaction between PVP and CNCs. For P80, the PVP chains entangled with PEO chains in amorphous region (**Figure 8a**). When 5 phr CNCs was introduced, the CNCs interacted with PVP, and helped free some PEO amorphous chains from the interaction with PVP. As a result, the freed PEO chains were rearranged into crystal lattice and therefore contributed to a higher crystallinity (**Figure 8b**) and the synergistic reinforcing effect. This assumption was verified by comparing the FTIR

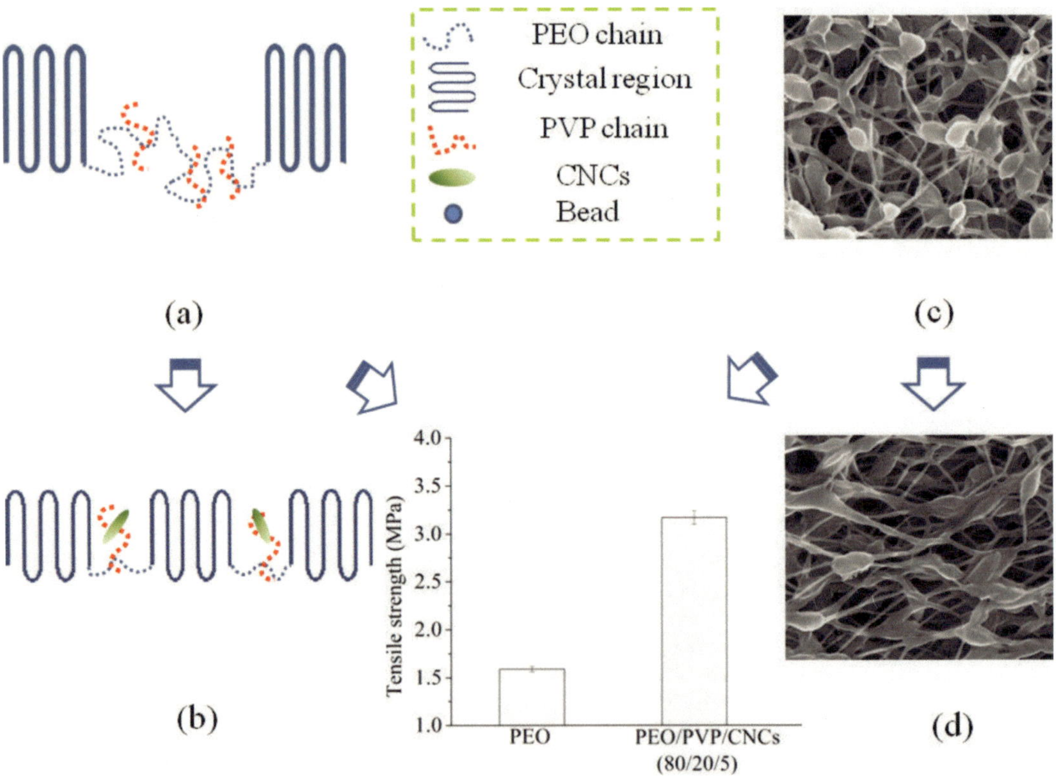

Figure 8. Schematic presentation of reinforcing mechanism: (a) P80 system; (b) P80C5 system; (c) fiber-bead-fiber morphology before tensile testing; (d) fiber-bead-fiber morphology after tensile testing.

results of P80 and P80C5, since the characteristic peak of carbonyl group of PVP was shifted to lower wave number, while all other characteristic peaks of PEO remained the same. In addition, the number of carbonyl group in PVP was considerably higher than the hydroxyl group (end group) in PEO. Therefore, CNCs were more prone to interact with PVP rather than PEO in general.

The second factor is related to the unique fiber-bead-fiber morphology of the prepared fiber mats. For P80C5, both the number and the size of the beads were the largest (**Figure 5**). And under the tensile loading, the beads were highly oriented and elongated along the tensile force direction (**Figure 8**). The elongation process of the beads (**Figure 8c** and **8d**) helped dissipate the applied energy. As a result, a surprising reinforcement was achieved. The third factor is due to the nature of CNCs and PVP. Both of them had a higher rigidity than that of PEO, which also accounted for the observed reinforcement.

It should be pointed out that the presence of spherical beads or elongated shape objects in electrospun fibers are often considered as defects in the manufactured materials [32, 33]. This could be due to the lower concentration of PEO, and the use of CNCs and PVP in the system. Further studies are needed to establish the reinforcing effect of combined CNCs and PVP in un-beaded electrospun fibers.

4. Conclusions

PVP and CNCs were used as reinforcement materials in electrospun PEO fiber mats. Addition of PVP and/or CNCs improved the overall thermal stability and mechanical properties of the PEO fibers. A synergistic reinforcing effect was observed when 20 phr PVP and 5 phr CNCs were used. Addition of CNCs induced special interactions with PVP and therefore disentangled the PVP and PEO chains. The freed PEO chains were rearranged into crystal lattice and contributed to a higher crystallinity as verified by the DSC results. In addition, the synergistic reinforcing effect was also accompanied with a fiber-bead-fiber morphology. Under the tensile loading, the beads were orientated and elongated, which helped dissipate more applied energy and therefore resulted in improved mechanical performance of the composites. This work demonstrates a new approach for reinforcing electrospun PEO-based composite fibers for sustainable green composite development.

Acknowledgements

We acknowledge the support from the Louisiana Board of Regents LEQSF(2017-18)-RD-A-01], Nanjing Forestry University (Nanjing, China), and Henan Academy of Science (Zhengzhou, China).

Author details

Qinglin Wu[1]*, Changtong Mei[2], Xiuqiang Zhang[3], Tingzhou Lei[3], Zhen Zhang[1] and Meichun Li[1]

*Address all correspondence to: wuqing@lsu.edu

1 School of Renewable Natural Resources, Louisiana State University, Baton Rouge, LA, USA

2 College of Materials Science and Engineering, Nanjing Forestry University, Jiangsu, China

3 Key Biomass Energy Laboratory of Henan Province, Zhengzhou, Henan, China

References

[1] Xu H, Xie L, Jiang X, Hakkarainen M, Chen J-B, Zhong G-J, Li Z-M. Structural basis for unique hierarchical cylindries induced by ultrahigh shear gradient in single natural fiber reinforced poly(lactic acid) green composites. Biomacromolecules. 2014;**15**(5):1676-1686. DOI: 10.1021/bm500100z

[2] Chen J, Lu L, Wu D, Yuan L, Zhang M, Hua J, Xu J. Green poly(e-caprolactone) composites reinforced with electrospun polylactide/poly(e-caprolactone) blend fiber mats. ACS Sustainable Chemistry & Engineering. 2014;**2**(9):2012-2110. DOI: 10.1021/sc500344n

[3] Sun Q, Mandalika A, Elder T, Nair SS, Meng X, Huang F, Ragauskas AJ. Nanocomposite film prepared by depositing xylan on cellulose nanowhiskers matrix. Green Chemistry. 2014;**16**(7):3458-3462. DOI: 10.1039/C4GC00793J

[4] Bodros E, Pillin I, Montrelay N, Baley C. Could biopolymers reinforced by randomly scattered flax fibre be used in structural applications? Composites Science and Technology. 2007;**67**(3-4):462-470

[5] Mohanty AK, Misra M, Drzal LT. Sustainable bio-composites from renewable resources: Opportunities and challenges in the green materials world. Journal of Polymers and the Environment. 2002;**10**(1-2):19-26. DOI: https://doi.org/10.1016/j.compscitech.2006.08.024

[6] Zhou C, Chu R, Wu R, Wu Q. Electrospun polyethylene oxide/cellulose nanocrystal composite nanofibrous mats with homogeneous and heterogeneous microstructures. Biomacromolecules. 2011;**12**(7):2617-2625. DOI: 10.1021/bm200401p

[7] Zhang Z, Zhao X, Zhang J, Chen S. Effect of nano-particles-induced phase inversion on largely improved impact toughness of PVC/α-methylstyrene–acrylonitrile copolymer (α-MSAN)/CPE-matrix composites. Composites Science and Technology. 2013;**86**:122-128. DOI: https://doi.org/10.1016/j.compscitech.2013.07.009

[8] Zhou C, Shi Q, Guo W, Terrell L, Qureshi AT, Hayes DJ, Wu Q. Electrospun bio-nanocomposite scaffolds for bone tissue engineering by cellulose nanocrystals reinforcing maleic anhydride grafted PLA. ACS Applied Materials & Interfaces. 2013;**5**(9):3847-3854. DOI: 10.1021/am4005072

[9] Zhang Z, Wu Q, Song K, Ren S, Lei T, Zhang Q. Using cellulose Nanocrystals as a sustainable additive to enhance hydrophilicity, mechanical and thermal properties of poly(vinylidene fluoride)/poly(methyl methacrylate) blend. ACS Sustainable Chemistry & Engineering. 2015;**3**(4):574-582. DOI: 10.1021/sc500792c

[10] Zhang Z, Wu Q, Song K, Lei T, Wu Y. Poly(vinylidene fluoride)/cellulose nanocrystals composites: Rheological, hydrophilicity, thermal and mechanical properties. Cellulose. 2015;**22**(4):2431-2441. DOI: 10.1007/s10570-015-0634-y

[11] Sun Q, Foston M, Sawada D, Pingali S, O'Neill H, Li H, Wyman C, Langan P, Pu Y, Ragauskas A. Comparison of changes in cellulose ultrastructure during different pretreatments of poplar. Cellulose. 2014;**21**(4):2419-2431. DOI: 10.1007/s10570-014-0303-6

[12] Azizi Samir MAS, Alloin F, Sanchez J-Y, Dufresne A. Cellulose nanocrystals reinforced poly(oxyethylene). Polymer. 2004;**45**(12):4149-4157. DOI: 10.1016/j.polymer.2004.03.094

[13] Xu X, Liu F, Jiang L, Zhu JY, Haagenson D, Wiesenborn DP. Cellulose nanocrystals vs. cellulose nanofibrils: A comparative study on their microstructures and effects as polymer reinforcing agents. ACS Applied Materials & Interfaces. 2013;**5**(8):2999-3009

[14] Xu X, Wang H, Jiang L, Wang X, Payne SA, Zhu JY, Li R. Comparison between cellulose nanocrystal and cellulose nanofibril reinforced poly(ethylene oxide) nanofibers and their novel shish-kebab-like crystalline structures. Macromolecules. 2014;**47**(10):3409-3416. DOI: 10.1021/am302624t

[15] Zhou Y, Qi P, Zhao Z, Liu Q, Li Z. Fabrication and characterization of fibrous HAP/PVP/PEO composites prepared by sol-electrospinning. RSC Advances. 2014;**4**(32):16731-16738. DOI: 10.1039/C3RA47168C

[16] Talebian S, Mehrali M, Mohan S, Balaji Raghavendran HR, Mehrali M, Khanlou HM, Kamarul T, Afifi AM, Abass AA. Chitosan (PEO)/bioactive glass hybrid nanofibers for bone tissue engineering. RSC Advances. 2014;**4**(90):49144-49152. DOI: 10.1039/C4RA06761D

[17] Cheng F, Gao J, Wang L, Hu X. Composite chitosan/poly(ethylene oxide) electrospun nanofibrous mats as novel wound dressing matrixes for the controlled release of drugs. Journal of Applied Polymer Science. 2015;**132**(24). DOI: 10.1002/app.42060

[18] Aliabadi M, Irani M, Ismaeili J, Piri H, Parnian MJ. Electrospun nanofiber membrane of PEO/chitosan for the adsorption of nickel, cadmium, lead and copper ions from aqueous solution. Chemical Engineering Journal. 2013;**220**:237-243. DOI: 10.1016/j.cej.2013.01.021

[19] Liu X, Xu Y, Wu Z, Chen H. Poly(N-vinylpyrrolidone)-modified surfaces for biomedical applications. Macromolecular Bioscience. 2013;**13**(2):147-154. DOI: 10.1002/mabi.201200269

[20] Cheng J, Wang S, Chen S, Zhang J, Wang X. Crystallization behavior and hydrophilicity of poly(vinylidene fluoride)/poly(methyl methacrylate)/poly(vinylpyrrolidone) ternary blends. Polymer International. 2012;**61**(3):477-484. DOI: 10.1002/pi.3185

[21] Pei A, Zhou Q, Berglund LA. Functionalized cellulose nanocrystals as biobased nucleation agents in poly(l-lactide) (PLLA)—Crystallization and mechanical property effects.

Composites Science and Technology. 2010;**70**(5):815-821. DOI: https://doi.org/10.1016/j.compscitech.2010.01.018

[22] Taylor L, Zografi G. Spectroscopic characterization of interactions between PVP and indomethacin in amorphous molecular dispersions. Pharmaceutical Research. 1997; **14**(12):1691-1698

[23] Song P, Xu Z, Guo Q. Bioinspired strategy to reinforce PVA with improved toughness and thermal properties via hydrogen-bond self-assembly. ACS Macro Letters. 2013;**2**(12):1100-1104. DOI: 10.1021/mz4005265

[24] Zhang Z, Wang S, Zhang J. Large stabilizing effect of titanium dioxide on photodegradation of PVC/α-methylstyrene-acrylonitrile copolymer/impact modifier-matrix composites. Polymer Composites. 2014;**35**(12):2365-2375. DOI: 10.1002/pc.22904

[25] Pielichowska K, Głowinkowski S, Lekki J, Biniaś D, Pielichowski K, Jenczyk J. PEO/fatty acid blends for thermal energy storage materials. Structural/morphological features and hydrogen interactions. European Polymer Journal. 2008;**44**(10):3344-3360

[26] Li J, Lai MF, Liu JJ. Control and development of crystallinity and morphology in poly(β-hydroxybutyrate-co-β-hydroxyvalerate)/poly(propylene carbonate) blends. Journal of Applied Polymer Science. 2005;**98**(3):1427-1436. DOI: 10.1002/app.22117

[27] Zhang Z, Song K, Li Y, Wu Q. Non-isothermal crystallization of poly (vinylidene fluoride)/poly (methyl methacrylate)/cellulose nanocrystal nanocomposites. International Journal of Polymer Analysis and Characterization. 2014;**19**(4):332-341. DOI: https://doi.org/10.1080/1023666X.2014.902530

[28] Zhang Z, Chen S, Zhang J, Li B, Jin X. Influence of chlorinated polyethylene on poly (vinyl chloride)/poly (α-methylstyrene-acrylonitrile) blends: Mechanical properties, morphology and thermal properties. Polymer Testing. 2010;**29**(8):995-1001. DOI: https://doi.org/10.1016/j.polymertesting.2010.09.003

[29] Zhang Z, Li B, Chen S, Zhang J, Jin X. Poly (vinyl chloride)/poly (α-methylstyrene–acrylonitrile)/acrylic resin ternary blends with enhanced toughness and heat resistance. Polymers for Advanced Technologies. 2012;**23**(3):336-342. DOI: 10.1002/pat.1876

[30] Fong H, Chun I, Reneker DH. Beaded nanofibers formed during electrospinning. Polymer. 1999;**40**(16):4585-4592. DOI: https://doi.org/10.1016/S0032-3861(99)00068-3

[31] Gupta P, Elkins C, Long TE, Wilkes GL. Electrospinning of linear homopolymers of poly(methyl methacrylate): Exploring relationships between fiber formation, viscosity, molecular weight and concentration in a good solvent. Polymer. 2005;**46**(13):4799-4810. DOI: https://doi.org/10.1016/j.polymer.2005.04.021

[32] Frenot A, Chronakis IS. Polymer nanofibers assembled by electrospinning. Current Opinion in Colloid & Interface Science. 2003;**8**(1):64-75. DOI: https://doi.org/10.1016/S1359-0294(03)00004-9

[33] Deitzel JM, Kleinmeyer J, Harris DEA, Tan NB. The effect of processing variables on the morphology of electrospun nanofibers and textiles. Polymer. 2001;**42**(1):261-272. DOI: 10.1016/S0032-3861(00)00250-0

Chapter 5

Functionalized Polyvinylidene Fluoride Electrospun Nanofibers and Applications

Dinesh Lolla, Lin Pan, Harshal Gade and
George G. Chase

Additional information is available at the end of the chapter

http://dx.doi.org/10.5772/intechopen.76261

Abstract

Electrospun polymeric nanofibers with flexible three-dimensional porous structures and high surface-to-volume ratio are potential resources for several novel applications in the fields of micro- and nanoscale filtration, water desalination, drug delivery, life sciences, catalysis, and energy harvesters. Functionalized polymeric fibers with enhanced molecular orientation, surface textural morphologies, and piezo-, pyro-, and ferroelectric properties are of technical and commercial interest around the world. Several emerging technologies including electrical polarization, vacuum plasma treatment, corona discharge, surface fluorination, and chemical treatments to functionalize the polyvinylidene fluoride nanofibers are discussed as potential applications of electroactive materials.

Keywords: electrospinning, polarization, aerosol filtration, salt absorption, catalysis

1. Polyvinylidene fluoride (PVDF) and its crystalline phases

Polyvinylidene fluoride (PVDF) is a semicrystalline, dielectric polymer with very high breakdown strength that offers long-duration surface charge retention, due to its unique dipole molecular structure with CH_2-CF_2 repeated monomer units [1]. PVDF is regarded as one of the most suitable polymeric materials to study polarizability in dielectric polymers. The dipole monomer structure of PVDF is favored in converting electromechanical coupling behavior with variance in thermomechanical processing. Although the piezoelectric coefficient of PVDF and its copolymers PVDF-HFP and PVDF-TrFE are less than piezoelectric ceramics like PBZ and $BaTiO_3$, their elasticity and mechanical stretchability make them more reliable materials for several emerging applications [2]. Moreover, PVDF exhibits higher piezoelectric response voltage, good thermal stability, and chemical resistance suitable for sensors, aerosol filters, actuators, fuel cells, energy harvesters, and other applications [3–7].

Depending on the crystalline conformations, PVDF exhibits five different molecular morphologies labeled as α, β, γ, δ, and ε. The composition and distinctive character of individual and binary phases are studied using Fourier transform infrared spectrometry (FTIR) and X-ray diffractometry (XRD) analysis. The α-phase consists of a non-centrosymmetric crystal structure with CH_2-CF_2 dipoles oriented in the same direction and has an all-trans (TTTT) planar zigzag conformation [8]. The β-phase follows a trans-gauche-trans-gauche′ (TGTG′) atomic arrangement of a centrosymmetric unit cell. In this atomic configuration, the CH_2 dipoles are perpendicular to CF_2 repeat units, and this produces a permanent electric dipole perpendicular to the axis with corresponding strong and ferroelectric and piezoelectric charges. The γ-phase has T_3GT_3G' conformations where the CH_2-CF_2 dipoles are oriented parallel to each other to form a non-centrosymmetric polar crystal. In general, the γ- and δ-phases form as a result of high-pressure crystallization [8, 9], are not commonly observed in the electrospun fibers, and hence are not considered further here.

Phase transformations among the crystal orientations take place under various postprocessing such as heat treatment, uniaxial stretching, and electrical poling [9–11]. Phase transformational mechanisms caused by polarization are interpreted in terms of vibrational motions around the individual atomic bonds in the PVDF molecule. Two distinct motions, known as flip-flop (segmental) and inversion motions (macromolecular), were observed during transformation of phases in polarization treatments. The segmental flip-flop motions usually occur at about 150°C (the Curie temperature) and result in a gradual change in the molecular conformations. Higher temperatures above 170°C are usually needed to produce inversion motions but can be achieved near the Curie temperature by subjecting the PVDF fibers with simultaneous electromechanical effects. Heat treatment in the presence of high electric fields and elevated ambient pressures can produce transformations from α- to β-phase and β- to γ-phase (<280°C, <4000 atm). The reverse transformations from β- to α-phase and γ- to α-phase typically require higher temperature and pressure (>290°C, >4500 atm). The transformation from β- to α-phase has so far only been studied in the unoriented state. The γ-phase PVDF melting temperature is about 15°C higher than the α- and β-phase materials. In this chapter, we reported several phase conversion techniques with primary focus on enhancing the amount of β-phase in PVDF fibers using different functionalization routes.

2. Electrospinning

Electrospinning is well documented and is considered an easy laboratory method for producing submicron and nanofibers [12, 13]. Electrospun polymer fibers are widely used in filtration [14, 15], catalysis [16–22], biomedical materials [23–26], and electrolytes [7, 27].

The application of electrical forces to produce polymer filaments began in the early 1930s. A brief summary of the early electrospinning literature is provided by Huang et al. [28]. A highly cited reference describing the mechanisms of electrospinning is by Reneker et al. [29].

The electrospinning process is driven by the electrical forces on the surface or inside the polymer solution. The free charges (ions) inside the polymer solution move in response to the electrical field and transfer a force to the solution [29]. When the electrically induced forces exceed the surface tension force, a liquid jet is ejected from the surface [30].

A schematic diagram of a typical laboratory electrospinning setup is shown in **Figure 1**. The components are high-voltage power supply, a syringe pump that delivers polymer solution through a tube to a small diameter needle, and a grounded rotating drum collector surface. The high voltage creates the electrical charge in the solution, and a jet is driven by the potential between the needle and the collector. As the jet travels to the collector, the solvent evaporates, and the jet solidifies into small fibers that deposit on the collector surface.

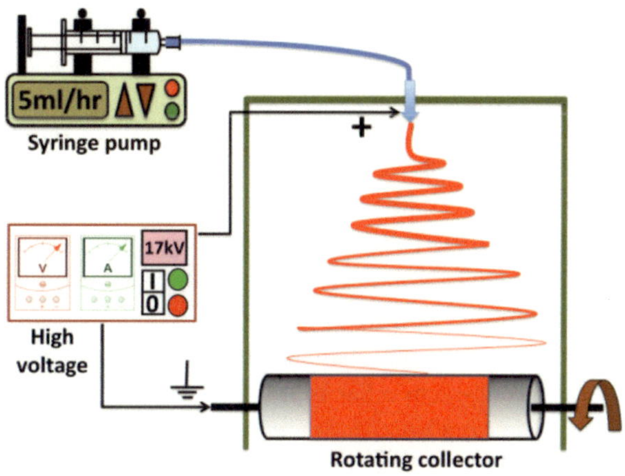

Figure 1. Schematic of a typical electrospinning setup used in this work.

MW	Conc. (%)	Solvent	Electrospun gap (volts)	Application	Ref.
107 K	20	DMF:DMA (1:1)	20 cm, 20 kV	Flat ribbons	[33]
	12–18	Ac:DMA (7:3)		Batteries	[31]
	25	DMA	15 cm, 10 kV	Electrolyte or separator	[37]
107 K		DMF:Ac (7:3) (v:v)	20 cm, 25 kV	Batteries	[34]
	15	DMF:Ac (2:8) (v:v)	8–15 kV	Metal cells	[38]
		DMF:Ac (6:4) (w:w)	12 cm, 25 kV 15 cm, 28 kV	Distillation	[39]
	20	DMF:Ac (7:3) (w:w)	15 cm, 25 kV	Separator	[40]
	14–24	DMF:Ac (3:7, 4:6, 5:5, 6:4, 7:3)	15 cm, 15–18 kV	Filtration	[41]
	16–20	DMS:Ac (1:1)	10–16 kV	Energy harvester	[42]
		DMF water (50:3) (w:w)	40 cm, 22.5 kV	Electrode	[43]

DMF, dimethylformamide; DMA, dimethylacetamide; DMS, dimethyl sulfoxide; Ac, acetone.

Table 1. Electrospinning of PVDF literature summary.

Many natural and synthetic polymers have been electrospun to produce fine fibers, such as polyacrylonitrile (PAN), polyvinyl alcohol (PVA), poly(methyl methacrylate) (PMMA), polyethylene oxide (PEO), polyethylene (PE), and polypropylene (PP) [31]. In comparison, PVDF has attracted much attention due to its properties and molecular structure. **Table 1** summarizes solution and spinning conditions for electrospinning PVDF fibers reported in literature. In addition, some researchers have modified the properties of the PVDF fiber mats by blending with other polymers [32]. As an example Gopalan et al. [34] mixed PVDF with varying amounts of PAN to fabricate fiber mats for use in lithium batteries. Ding and coworkers mixed PVDF with PMMA for same purpose, Guo et al. [35] prepared the PU/PVDF electrospun scaffolds for wound healing, and Dong et al. [36] electrospun PVDF/PTFE membranes for distillation.

3. Functionalization of polyvinylidene fluoride nanofibers

Properties of nanofibers, such as electrical, textural, optical, adhesive, and tensile strength, are highly dependent on the inherent polymeric properties and internal molecular structure. The chemical modification of polymer nanofibers introduces new characteristics to the materials that extend and enhance the scope of their industrial applications over several orders of magnitude. Hence, enhancement of molecular orientation of the PVDF nanofibers has attracted interest of the scientific community. A number of functionalization techniques are available in pilot and commercial scale operations. These functionalization processes can be economical, profitable, environmentally friendly, and long-term reliable [44, 45].

Fiber stretching during the electrospinning process causes dipoles to align perpendicular relative to each other [44]. Piezoresponse force microscopy (PFM) was used to analyze piezoelectric responses and ferroelectric domains in individual electrospun nanofibers with diameters 70, 170, and 400 nm [45]. The β-phase compositions of individual fibers were estimated in the 80–87% range using Beer-Lambert's law, confirming that fibers with smaller diameters experienced higher oriented conformational changes consistent with stronger elongational forces such as those produced with near-field electrospinning (NFES) due to short tip to collector distances.

Liu et al. [46] studied processing and solution conditions to obtain the highest β-content when PVDF was electrospun together with multiwalled carbon nanotubes (MWCNTs). Distinct oriented crystalline structures of the MWCNT/PVDF in aligned nanofibers were obtained. Due to the nucleation of highly oriented fibers and extended molecular crystallites at the interface, NFES techniques showed 28% increase in β-phase with 0.05% wt% of MWCNTs.

Served et al. [11] subjected a pre-stretched 100-μm-thick PVDF film containing exclusively α-phase with 5% head-to-head- and tail-to-tail-type (HHTT) defects to electrical poling at 80 and 170°C (T_m-178°C). Aluminum electrodes were placed on either side of the film, and a DC electric field of 1 MV/cm was used for charging. At 170°C the film changed from nonpolar α-phase to polar β-phase with TGTG molecular conformations. The β-phase polarized films showed a strong piezoelectric coefficient, d_{33} = 8.5pC/N, and was validated with XRD and FTIR analysis. Salimi et al. [47] analyzed β-content in compression-molded PVDF films made of two different grades of raw polymer (Kynar® 720, Hylar® MP10). A maximum of 74% β-phase was observed for films that were 38–40% crystalline at 90°C and stretched at a ratio between 4.5 and 5. A similar study of nanoscale domain imaging and spatial distribution of d_{33} on single electrospun fibers concluded that the d_{33} distribution was more uniform along the length of the fibers compared to cross-sectional diameter [48].

Real-time piezoelectric responses were observed by manipulating the operational voltages of a PFM at a spring constant of 0.11 Nm^{-1}, with cantilever resonating frequency of 135 kHz and the amplitude changing stepwise from −30 to +30 V. The highest deflection in the piezoresponse hysteresis loop at 3.3 nm was observed at a V_{dc} of -30 V. XRD patterns decrypted using a curve deconvolution technique revealed 72.7% β-phase at plane (110,200) and 15.1% α-phase (202) and indicated voltages as much as ±30 V can cause significant effects on nanoscale β-phase nanocrystals. Nanodomains were distributed along to the fiber axis, and the β-phase orientation was investigated by TEM and XRD [48]. Furthermore, polymer composites can be polarized at low electric field strengths with addition of nanoparticle ferroelectric ceramics [49, 50].

Introduction of chemical functional groups into virgin PVDF polymer has resulted in novel functional characteristics [51, 52]. Several studies have reported the ability of modifying structural morphologies and analytical properties in electrospun PVDF nanofibers via plasma deposition of polymers under inert conditions. Molecular cross-linking on PVDF can be achieved through dehydrofluorination or by introducing functional comonomers during electrospinning [5, 8, 53].

The PVDF materials with modified properties are of significant practical interest. In filtration, for example, an exceptional particle capture efficiency of ≥99.999% was achieved by a hybrid monolithic electret aerogel composed of syndiotactic polystyrene (sPS)/PVDF [54], whereas 98.9% filtration efficiency was recorded with sPS monolithic aerogel comprised of similar solid content. In comparison, performance of cellulose acetate electrospun fibers with diameters in the range from 0.1 to 24 μm challenged with a solid brine aerosol (NaCl) and a liquid aerosol of (diethyl hexyl sebacate) showed a maximum efficiency of 70% and with the most penetrating particle size in the range of 40–270 nm [55].

4. Results and discussion

4.1. Electrospinning of PVDF fibers

Electrospinning solutions 10 wt% were prepared by dissolving Kynar® 761 grade resin (MW of about 550,000, melt viscosity of 35 kp, and a solution viscosity of 350cp at room temperature), PVDF powder (Arkema Inc., USA) in cosolvents N-N-dimethylformamide (DMF), and acetone (Sigma-Aldrich, USA). The solutions were electrospun under the conditions reported in **Table 2**.

Mats of 20 g/m^2 basis weight were preheated in an oven at 70°C for 4 hrs before any analysis. Fiber morphologies were analyzed under a scanning electron microscope. Smooth and consistent fibers were observed as shown in **Figure 2**. A maximum of 57.7% β-content in the fibers was observed. SEM images occasionally showed branched fibers. Branched fibers were formed because of "static equilibrium undulations under the combined effect of the electric Maxwell stresses and surface tension as the electrical stresses are increased" [56].

Conc. PVDF (wt%)	DMF: acetone (w: w)	Gap distance (cm)	Voltage (kV)	Flow rate (mL/h)	Collector rotation (RPM)	Avg. fiber dia. (nm)	Standard deviation (nm)
10	1:1	20	17	5	100	196	54

Table 2. Electrospinning conditions and average PVDF fiber diameter.

Figure 2. SEM image of 10 wt% PVDF, 1:1 DMF/acetone electrospun fibers.

4.2. Atomic resolution electron microscopy of PVDF nanofibers

A segment of 8X5nm PVDF fiber was studied under an aberration-corrected electron microscope with highly controlled electron beam shown in **Figure 3**. The images revealed the paths of individual monomers aligned in the direction of fiber axis as shown in **Figure 3A**. CF_2 bonds appeared as brighter dots compared to other bonds as gray and black dots. The raw TEM micrograph was converted into a Fourier transform image to reduce the electron noise and reverted as an RGB image to enhance the features. Paths of CF_2 molecules from end to end are clearly seen in the enhanced RBG image in **Figure 3B**. The calculated distance between the centers of two adjacent bright dots in **Figure 3B** is about 0.25 nm which is consistent with molecular dynamic simulation of the theoretical distance between fluorine atoms in the β-phase crystallographic structure of the PVDF [1] in **Figure 3C**.

4.3. Functionalization of PVDF nanofibers by electrical polarization

Lolla et al. [7] describe a thermal-stretch-electric field polarization treatment of PVDF nanofibers to fabricate polarized PVDF fiber mats. Simultaneous thermal and electrical treatments caused substantial changes in surface textural morphology. These surface morphological changes are obvious when as-spun fibers shown in **Figure 4A** are compared to the thermal-electrically treated fibers in **Figure 4B**.

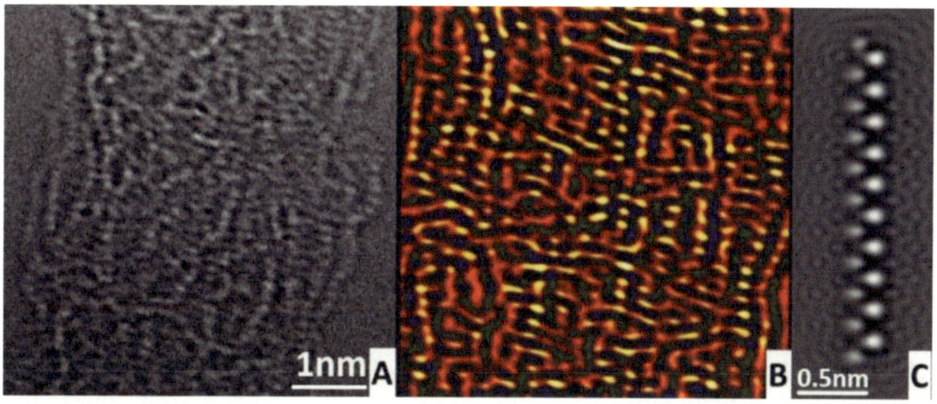

Figure 3. (A) Raw high magnification TEM image of PVDF nanofiber, (B) Fourier transform of raw image, and (C) molecular dynamic simulation of PVDF molecule.

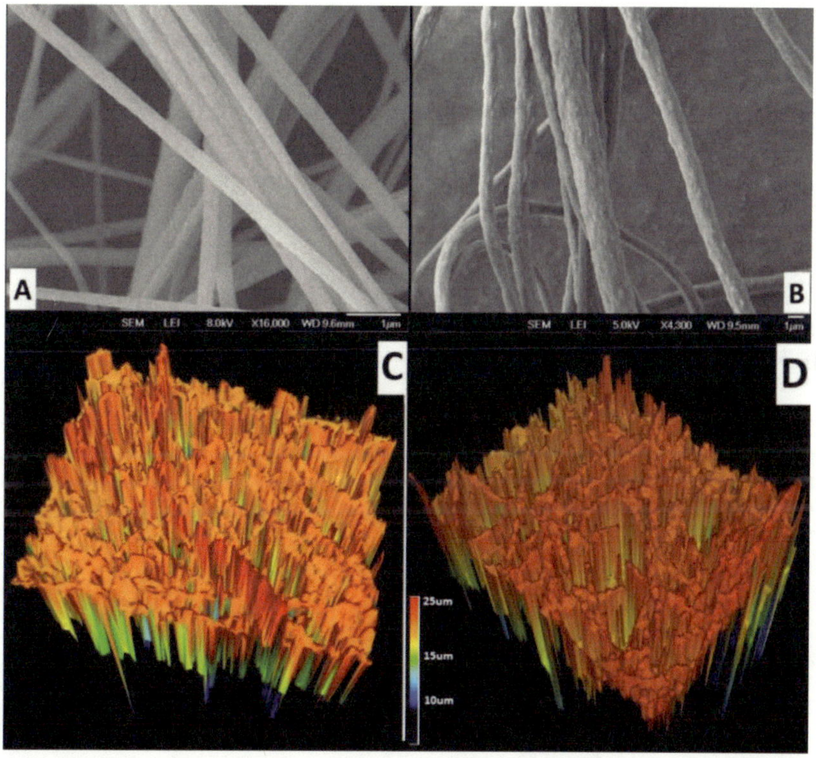

Figure 4. Surface morphology analysis using high magnification SEM (images A and B) and laser microscopy (images C and D). The fiber samples in images A and C were as-spun fibers, and the images B and D were thermal-electric treated polarized PVDF nanofibers.

Surface modifications were found to have a remarkable effect during liquid–liquid filtration applications compared to aerosol filtration as the interfacial strength between liquid droplets is much higher in polarized fiber mats compared to relatively smooth fibers. Kinetic energy generated by electron collision during charge migration is suspected to be the primary reason for surface irregularities. SEM analysis provided only 2D visual conformation of polarization-associated surface modifications. A complete three-dimensional analysis was done to obtain precise increase in roughness due to electron interaction. Lasers were projected in z-direction with the fibers across a 20×30×25μm sample, and few thousands of data points were gathered from hundreds of fibers to make the analysis. All the fiber samples were highly irregular with several mounds, hills, and valley-like structures as shown in **Figure 4C** and **D**. Unlike the 2D SEM images, the detailed layer-by-layer fiber interactions were seen, and paths of fiber conglutination or fiber cross-linking were more accurately captured in several directions using laser projections.

2D images of laser intensities were obtained in parallel with three-dimensional images, and an empirical analysis was conducted to estimate the changes in surface morphology. As-spun fibers and thermal-electric treated fibers are shown in **Figure 5A** and **B**. The data analysis was conducted on the circular areas highlighted in the images. The radii of the circles in both images were 40 μm. The average intensities of as-spun fibers from peak to valley detection were mapped as shown in **Figure 5C** represented by the blue line, which gave us a mean surface roughness of R_{ms} = 7.86 ± 4.73 nm. Similar calculations were also performed on the thermal-electric treated fibers, and the mean surface roughness R_{ms} of 16.86 ± 6.68 nm was

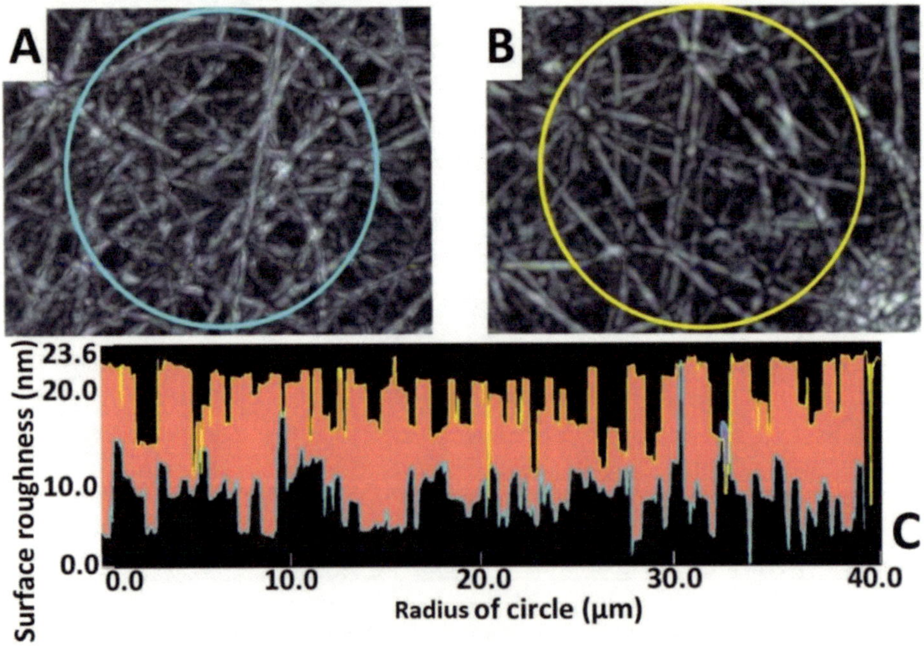

Figure 5. Comparison of surface roughness (A) as spun (B) polarized fibers (C) graphical overlay of surface roughness distribution.

determined which proves a substantial rise in surface roughness. Average values of roughness averaged over the length of the circumference of circles of varying radius from 0 to 40 μm was used to make a comparison between as-spun and polarized fibers.

4.3.1. Nanoscale aerosol particle filtration

Functionalized electrospun fibers are of great interest in aerosol filtration [57]. Fiber mats were subjected to aerosols of 10–250 nm diameter NaCl particles using a TSI-automated filter tester (TSI 8130). Each test was conducted for a duration of 10 s at 10 l/min volumetric flowrate. Three individual samples were consecutively tested 30 times with 30 days between tests to generate a particle capture v/s pressure buildup profile. For thermal-electrically treated polarized fiber mats, the first tests were performed within 24 h of polarization. Both the as-spun and polarized samples showed very distinct and diverse capture trends as apparent from **Figure 6A** and **B**. Further research with these materials and with theoretical predictions is needed to explore and understand the shelf life of the filter media in association with net charge. The polarized fibers did not exhibit cake formation, even for the smallest fiber diameters, and had much smaller pressure drop compared to the as-spun fibers. Almost all of the aerosol particles were evenly distributed among individual fibers in the polarized mat as compared to the agglomerates observed in the mat of as-spun fibers.

The plots in **Figure 7A** show the filter efficiency and pressure drop as a function of the number of tests. Effectively, the plot shows the filter performance over time as it was affected by loading of particles and by charge dissipation (if dissipation occurs) over an extended time. Both the as-spun and polarized filters recorded similar efficiencies of 94.63 ± 012 and 94.96 ± 0.46 during the first experimental run with pressure drops of 56 ± 1.63 and 49.66 ± 1.69 mmH_2O, respectively. Pressure drops across the media are in good agreement with the air permeability as shown in **Figure 7B**.

The Frasier air permeabilities of the fiber mats were tested at two different test pressures at 125 and 2000 kPa. The Darcy law permeability has units of area, whereas the Frazier permeability is reported as volumetric flow rate (cfm = cubic feet per minute). The Darcy permeability can be calculated, but for the purposes here, the relative magnitudes of the two flow rates are the relevant data. The plot in **Figure 7B** shows that the relative flow rates of the polarized mats

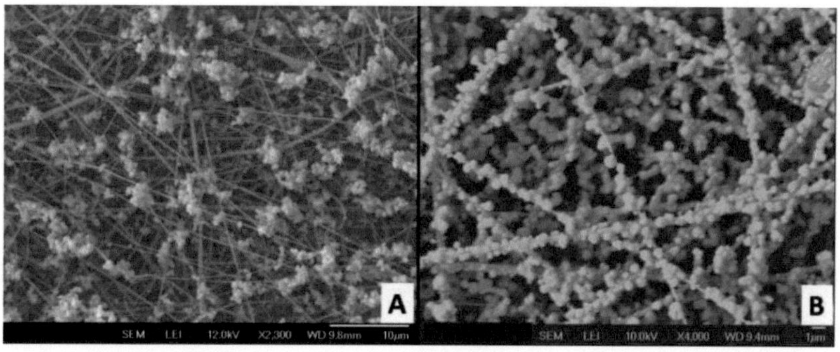

Figure 6. Brine (NaCl) aerosol captures on (A) as-spun and (B) polarized PVDF filter media.

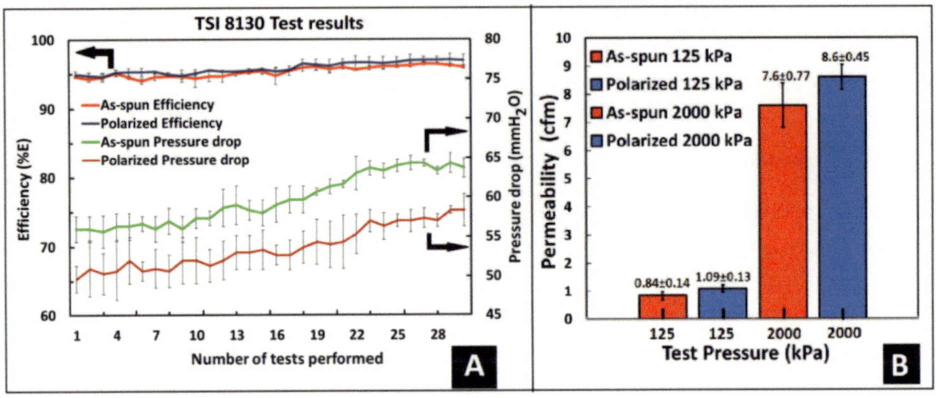

Figure 7. Aerosol penetration testing and Frazier permeability (i.e., flow rate at applied pressure) of as-spun and polarized fibers.

were about 17% greater than the flow rates of the as-spun mats, which corresponds to about 17% greater Darcy permeabilities in the polarized mats.

The pressure of 2000 psi is 16 times the pressure drop of 125. If the Darcy permeability was constant, then one would expect the flow rate to increase by a factor of 16 as the pressure increased. The data show that an increase of flow rate was only on order of a factor of 8 times. This indicates that as the flow rate increased the fiber mat structures may have deformed and caused a higher resistance to flow. This topic needs future investigation.

Inspection of SEM images showed attraction between fibers in the polarized mat that caused the fibers to rearrange relative to each other in the fiber mat which resulted in larger pores than the pores in the as-spun mat and is a likely cause of the increase in permeability of the polarized mats. At the end of the 30 filtration experiments, a slight increase in efficiency due to particle accumulation was observed in both the samples. The as-spun fiber mats had a maximum efficiency of 96% at 64 mmH_2O pressure drop, and the polarized fibers had a maximum efficiency of 97% pressure drop of 58 mmH_2O. Because both efficiencies were very similar, the significant advantage of the polarized mats was the reduced pressure drop.

4.3.2. Functionalized PVDF nanofibers in water desalination and purification

Population growth, industrialization, rise in living standards, and rapid climate changes have an increased demand for water significantly [58]. Water desalination and purification are a possible solution for providing fresh drinking water to the world especially in drought-prone regions [59]. Researchers have developed several treatment processes such as reverse osmosis (RO), nanofiltration (NF), ultrafiltration (UF), and thermal methods such as membrane distillation to improve water quality and supply. These techniques are energy intensive and have high operating and maintenance costs which make it difficult for developing countries to implement [59]. Membrane distillation finds limited application due to lack of a variety of membranes that can produce stable and high flux for a long time [60].

Industrial effluents contain a wide range of hazardous and toxic substances including heavy metal ions (Cd^{2+}, Pb^{2+}, Hg^{2+}, Zn^{2+}, etc.), organic acids, nitro compounds, hydrocarbons, sulfides (S^{2-}), sulfites (SO_3^{2-}), and sugars. Heavy metal pollution can cause serious environmental and health problems to humans [61]. Various methods used for heavy metal removal include ion exchange, electrodialysis, chemical precipitation, and solid-phase extraction [62–64]. Materials such as nanoparticles, polymers, and organic and inorganic compounds have been employed in the form of thin films, membranes, or powder for water treatment [65]. Apart from these, using a nano-adsorbent for heavy metal removal via adsorption mechanisms is a growing area of research because of its large surface area and mechanical strength [66]. However, regeneration of nano-adsorbents after water treatment is a challenge as adsorption activity decreases with time due to agglomeration. To overcome this challenge, nano-adsorbents can be modified using functionalization techniques [67].

Blending PVDF with inorganic materials such as ZrO_2 [68], ZnO [69], Al_2O_3 [70], Fe_3O_4 [71], CdS [72], SiO_2 [73], and TiO_2 [74] to increase adsorption capacity can help in heavy metal ion removal. This research area is of growing interest. For example, Zhang et al. [75] used ZnO-hybridized (PVDF/ZnO) membranes for adsorption and desorption studies of Cu^{2+} ions.

Zhao et al. [76] studied melamine-diethylenetriaminepentaacetic acid/polyvinylidene fluoride (MA-DTPA/PVDF)-chelating membranes bearing polyaminecarboxylate groups for removal of Ni^{2+} ions from wastewater. Salehi et al. [77] studied adsorption of Ni^{2+} and Cd^{2+} ions using 8-hydroxyquinoline ligand-immobilized PVDF membrane.

Na^{+}, Cl^{-}, and SO_4^{2-} ions are present in significant concentrations in typical seawater and brackish waters [78]. Most of the feeds subjected to desalination processes have sodium chloride (NaCl) or sulfates of Ca and Mg.

Table 3 is a brief literature summary of the electrospun fiber membranes applied to desalination performance. PVDF is generally applied in the as-spun condition. Few data are available on performance of functionalized PVDF for this application.

4.3.3. Membrane and polymer nanofiber catalyst

Membrane-based separations and heterogeneous chemical reactions are often treated as independent processes. The advantages of combining the two operations have drawn attention to membrane reactors that combine reaction and separation in a single-unit operation [86]. The properties of PVDF fiber mats naturally lend themselves to use as membrane reactors. The PVDF fiber mats have strength, can be embedded or coated with catalyst particles, have thermal stability over a useful temperature range, are inert to many chemical environments, can be superhydrophobic, and thus provide a barrier to aqueous solutions while being porous to gases.

Catalytic membrane reactors can be fabricated of materials that can selectively remove the reaction products from the reactor to increase the product yield. Membranes as catalyst support structures can provide relatively large surface areas, especially when the electrospun fibers are very small, for supporting catalyst particles [87].

Electrospun layer	Second layer/ treatment	Solute	Method	Flux ($L/m^2/h$)	Rejection	Ref.
PVDF	Polyamides	$MgSO_4$	TFNC by interfacial	0.66	75.7	[79]
		NaCl		0.66	70.2	
PVDF	n.a.	6%wt NaCl	AGMD	11–12 kg/m^2 h	n.a.	[80]
PVDF, clay nanocomposites	n.a.	NaCl	DCMD	n.a.	98.27 99.95	[81]
PET/PS	Polyamides	NaCl	Interfacial	1.13 $L\ m^{-2}h^{-1}bar^{-1}$	n.a.	[82]
PVDF-HFP (hot pressed)	Hot pressed	NaCl	DCMD	20–22 $L\ h^{-1}\ min^{-2}$	98	[83]
PAN	Polyamides	$MgSO_4$	TFNC interfacial	81	84.5	[84]
PVDF-PTFE	Microporous PTFE	NaCl	VMD	18.5 kg/m^2 h	99.9	[36]
PVDF-co-HFP	PAN microfibers	35 g/L NaCl	DCMD	45–30 $L\ h^{-1}\ min^{-2}$	n.a.	[85]
Intrinsically modified PVDF	Ag nanoparticles	3.5 wt% NaCl	DCMD	31.8 $L\ h^{-1}\ min^{-2}$	n.a.	[60]

DCMD, direct contact membrane distillation; AGMD, air gap membrane distillation; TFNC, thin-film nanocomposite; VMD, vacuum membrane distillation.

Table 3. Performance of various electrospun polymeric nanofibers used in water desalination techniques in pristine form or in modified conditions.

Inorganic membranes can provide high-temperature durability and easy loading of catalyst [88]. However, polymer membranes have the advantages including flexibility, easy for recycling [89], and affinity for reagents [90].

Electrospun fiber membranes have been studied for their physical and chemical properties, mechanical performance [28], large surface areas, and high porosities [19]. Pinto et al. [16] studied polystyrene electrospun fibers for catalysts and nanopore filter applications. Electrospun PVDF nanofibers were studied by Li et al. [19] for immobilizing $CoCl_2$ catalyst for hydrolysis of $NaBH_4$. The high thermal stability, moduli, and mechanical strength of the PVDF fibers showed excellent catalytic activity and recycling stability.

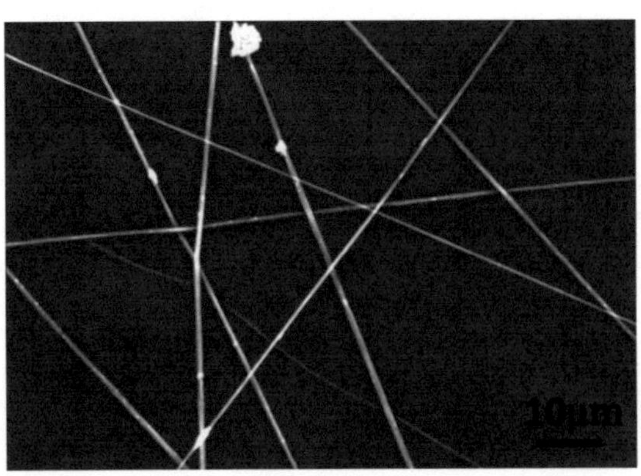

Figure 8. SEM image of electrospun PVDF+Pd black fibers.

In the work here, palladium (Pd) immobilized on PVDF electrospun fiber mats was investigated for catalytic hydrogenation of phenol to cyclohexanone. The one-step reaction can directly hydrogenate phenol into cyclohexanone, and the hydrogenation can be conducted either in liquid or gas phase. A two-step reaction is also possible in which phenol is first hydrogenated to cyclohexanol and then dehydrogenated to cyclohexanone [91]. PVDF and PVDF-HFP electrospun fiber mats are hydrophobic, resist the flow of water through the membrane, and provide a barrier between phenol water solution and hydrogen gas. **Figure 8** has SEM images of PVDF+5% Pd black samples.

Figure 9 shows EDX images of PVDF fibers with Pd black particles. The elemental Pd (appears in green color) on fibers. The fibers appear red due to elemental fluoride. Similar results were obtained for PVDF-HFP electrospun fiber mats.

Batch tests were conducted with Pd supported on PVDF-HFP fibers with 5, 10, and 15 wt% of Pd black. The average fiber diameter was 357 nm. The fiber mats were immersed in 75 mL of phenol/water solution (20 g/L) at 80°C under mild stirring and exposed to H_2 gas bubbles.

Reaction sample concentrations were measured by GC. The conversion and selectivity were calculated based on concentration changes. The reaction conversion increased with the concentration of Pd black and reached 98% conversion after 7 hours. The selectivity for cyclohexanone was about 97% for all of the fiber samples.

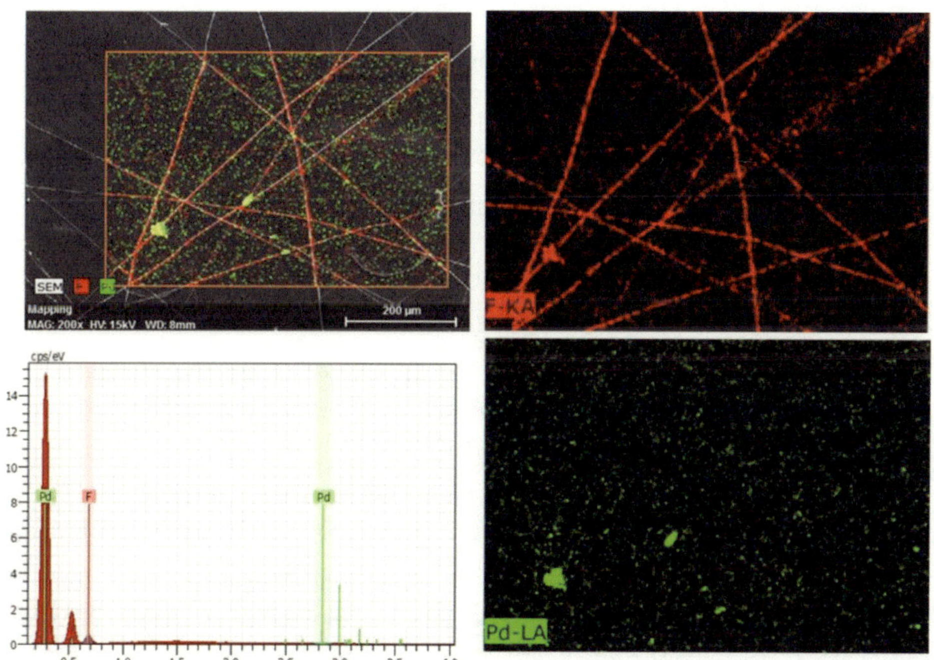

Figure 9. Energy dispersity X-ray (EDX) images of electrospun PVDF +5 wt% Pd black fibers.

5. Conclusion

In this work, PVDF and related copolymer mixtures are discussed. PVDF has unique properties due to its CH_2-CF_2 repeated monomer units that make it a material of recent scientific interest. Several applications of the electrospun PVDF polymer were reviewed. The PVDF molecule can be polarized. The polarized fiber mats were tested as aerosol filter media. SEM images showed remarkably different performances due to changes in particle capture mechanisms.

The PVDF membranes have potential applications for water treatment, first as a filter but second as a desalination membrane. The inherent dipole charges due to the CH_2-CF_2 repeated monomer units may be useful for separating salt ions from water. The last topic discussed is the use of PVDF and related copolymers as catalyst supports. As an example, experimental data for hydrogenation of phenol is presented. The limited amount of experimental data available showed that the PVDF membranes can be used for these applications. Further work is needed on these topics to determine the full potential of PVDF and related copolymer electrospun fiber mats.

Author details

Dinesh Lolla[1], Lin Pan[2]*, Harshal Gade[2] and George G. Chase[2]

*Address all correspondence to: lp57@zips.uakron.edu

1 Bioscience and Water Filtration Division, Parker-Hannifin Corporation, Oxnard, CA, United States

2 Department of Chemical and Biomolecular Engineering, The University of Akron, Akron, OH, United States

References

[1] Lolla D, Gorse J, Kisielowski C, Miao J, Taylor PL, Chase GG, et al. Polyvinylidene fluoride molecules in nanofibers, imaged at atomic scale by aberration corrected electron microscopy. Nanoscale. 2016;**8**:120-128. DOI: 10.1039/C5NR01619C

[2] Kilic A, Shim E, Yeom BY, Pourdeyhimi B. Improving electret properties of PP filaments with barium titanate. Journal of Electrostatics. 2013;**71**:41-47. DOI: 10.1016/j.elstat.2012.11.005

[3] Damaraju SM, Wu S, Jaffe M, Arinzeh TL. Structural changes in PVDF fibers due to electrospinning and its effect on biological function. Biomedical Materials. 2013;**8**:045007. DOI: 10.1088/1748-6041/8/4/045007

[4] Gaur A, Kumar C, Shukla R, Maiti P. Induced piezoelectricity in poly(vinylidene fluoride) hybrid as efficient energy harvester. ChemistrySelect. 2017;**2**:8278-8287. DOI: 10.1002/slct.201701780

[5] Li H-Y, Liu Y-L. Nafion-functionalized electrospun poly (vinylidene fluoride)(PVDF) nanofibers for high performance proton exchange membranes in fuel cells. Journal of Materials Chemistry A. 2014;**2**:3783-3793. DOI: 10.1039/C3TA14264G

[6] Liu S, Kim JT, Kim S. Effect of polymer surface modification on polymer–protein interaction via hydrophilic polymer grafting. Journal of Food Science. 2008;**73**. DOI: 10.1111/j.1750-3841.2008.00699.x

[7] Lolla D, Lolla M, Abutaleb A, Shin HU, Reneker DH, Chase GG. Fabrication, polarization of electrospun polyvinylidene fluoride electret fibers and effect on capturing nanoscale solid aerosols. Materials. 2016;**9**:671. DOI: 10.3390/ma9080671

[8] Sessler G. Piezoelectricity in polyvinylidenefluoride. The Journal of the Acoustical Society of America. 1981;**70**:1596-1608. DOI: 10.1121/1.387225

[9] Baniasadi M, Xu Z, Cai J, Daryadel S, Quevedo-Lopez M, Naraghi M, et al. Correlation of annealing temperature, morphology, and electro-mechanical properties of electrospun piezoelectric nanofibers. Polymer. 2017;**127**:192-202. DOI: 10.1016/j.polymer.2017.08.053

[10] Tashiro K, Takano K, Kobayashi M, Chatani Y, Tadokoro H. Phase transition at a temperature immediately below the melting point of poly (vinylidene fluoride) from I: A proposition for the ferroelectric curie point. Polymer. 1983;**24**:199-204. DOI: 10.1016/0032-3861(83)90133-7

[11] Servet B, Rault J. Polymorphism of poly (vinylidene fluoride) induced by poling and annealing. Journal de Physique. 1979;**40**:1145-1148. DOI: 10.1051/jphys:0197900400120114500

[12] Chase G, Zhang X. Solid aerosol filtration by electrospun poly vinyl pyrrolidone fiber mats and dependence on pore size. Journal of Textile Engineering and Fashion Technology. 2017;**1**. DOI: 10.15406/jteft.2017.01.00030

[13] Zhang X, Chase GG. Electrospun elastic acrylonitrile butadiene copolymer fibers. Polymer. 2016;**97**:440-448. DOI: 10.1016/j.polymer.2016.05.063

[14] Shin C, Chase GG, Reneker DH. Recycled expanded polystyrene nanofibers applied in filter media. Colloids and Surfaces A: Physicochemical and Engineering Aspects. 2005;**262**:211-215. DOI: 10.1016/j.colsurfa.2005.04.034

[15] Viswanadam G, Chase GG. Water–diesel secondary dispersion separation using superhydrophobic tubes of nanofibers. Separation and Purification Technology. 2013;**104**:81-88. DOI: 10.1016/j.seppur.2012.11.020

[16] Torres B, editor. Ultrafine fibers of polystyrene dissolved in tetrahydrofuran prepared using the electrospinning method. In: Proceeding of the National Conference on Undergraduate Research; 2001 15-17 March 2001; Kentucky

[17] Cheng HH, Chen F, Yu J, Guo ZX. Gold-nanoparticle-decorated thermoplastic polyurethane electrospun fibers prepared through a chitosan linkage for catalytic applications. Journal of Applied Polymer Science. 2017;**134**. DOI: 10.1002/app.44336

[18] Shi W, Li H, Zhou R, Qin X, Zhang H, Su Y, et al. Preparation and characterization of phosphotungstic acid/PVA nanofiber composite catalytic membranes via electrospinning for biodiesel production. Fuel. 2016;**180**:759-766. DOI: 10.1016/j.fuel.2016.04.066

[19] Li Q, Chen Y, Lee DJ, Li F, Kim H. Preparation of Y-zeolite/$CoCl_2$ doped PVDF composite nanofiber and its application in hydrogen production. Energy. 2012;**38**:144-150. DOI: 10.1016/j.energy.2011.12.021

[20] Shin HU, Lolla D, Nikolov Z, Chase GG. Pd–Au nanoparticles supported by TiO_2 fibers for catalytic NO decomposition by CO. Journal of Industrial and Engineering Chemistry. 2016;**33**:91-98. DOI: 10.1016/j.jiec.2015.09.020

[21] Shin HU, Abutaleb A, Lolla D, Chase GG. Effect of calcination temperature on NO–CO decomposition by Pd catalyst nanoparticles supported on alumina nanofibers. Fibers. 2017;**5**:22. DOI: 10.3390/fib5020022

[22] Abutaleb A, Lolla D, Aljuhani A, Shin HU, Rajala JW, Chase GG. Effects of surfactants on the morphology and properties of electrospun polyetherimide fibers. Fibers. 2017;**5**:33. DOI: 10.3390/fib5030033

[23] Kenawy E-R, Bowlin GL, Mansfield K, Layman J, Simpson DG, Sanders EH, et al. Release of tetracycline hydrochloride from electrospun poly (ethylene-co-vinylacetate), poly (lactic acid), and a blend. Journal of Controlled Release. 2002;**81**:57-64. DOI: 10.1016/S0168-3659(02)00041-X

[24] Matthews JA, Wnek GE, Simpson DG, Bowlin GL. Electrospinning of collagen nanofibers. Biomacromolecules. 2002;**3**:232-238. DOI: 10.1021/bm015533u

[25] Kenawy E-R, Layman JM, Watkins JR, Bowlin GL, Matthews JA, Simpson DG, et al. Electrospinning of poly (ethylene-co-vinyl alcohol) fibers. Biomaterials. 2003;**24**:907-913. DOI: 10.1016/S0142-9612(02)00422-2

[26] Rajala JW, Shin HU, Lolla D, Chase GG. Core–shell electrospun hollow aluminum oxide ceramic fibers. Fibers. 2015;**3**:450-462. DOI: 10.3390/fib3040450

[27] Choi SW, Kim JR, Ahn YR, Jo SM, Cairns EJ. Characterization of electrospun PVDF fiber-based polymer electrolytes. Chemistry of Materials. 2007;**19**:104-115. DOI: 10.1021/cm060223+

[28] Huang Z-M, Zhang Y-Z, Kotaki M, Ramakrishna S. A review on polymer nanofibers by electrospinning and their applications in nanocomposites. Composites Science and Technology. 2003;**63**:2223-2253. DOI: 10.1016/S0266-3538(03)00178-7

[29] Reneker DH, Yarin AL, Fong H, Koombhongse S. Bending instability of electrically charged liquid jets of polymer solutions in electrospinning. Journal of Applied Physics. 2000;**87**:4531. DOI: 10.1063/1.373532

[30] Katta P, Alessandro M, Ramsier R, Chase G. Continuous electrospinning of aligned polymer nanofibers onto a wire drum collector. Nano Letters. 2004;**4**:2215-2218. DOI: 10.1021/nl0486158

[31] Kim JR, Choi SW, Jo SM, Lee WS, Kim BC. Electrospun PVdF-based fibrous polymer electrolytes for lithium ion polymer batteries. Electrochimica Acta. 2004;**50**:69-75. DOI: 10.1016/j.electacta.2004.07.014

[32] Kang G-D, Cao Y-M. Application and modification of poly (vinylidene fluoride)(PVDF) membranes—A review. Journal of Membrane Science. 2014;**463**:145-165. DOI: 10.1016/j.memsci.2014.03.055

[33] Koombhongse S, Liu W, Reneker DH. Flat polymer ribbons and other shapes by electrospinning. Journal of Polymer Science Part B: Polymer Physics. 2001;**39**:2598-2606. DOI: 10.1002/polb.10015

[34] Gopalan AI, Santhosh P, Manesh KM, Nho JH, Kim SH, Hwang C-G, et al. Development of electrospun PVdF–PAN membrane-based polymer electrolytes for lithium batteries. Journal of Membrane Science. 2008;**325**:683-690. DOI: 10.1016/j.memsci.2008.08.047

[35] Guo H-F, Li Z-S, Dong S-W, Chen W-J, Deng L, Wang Y-F, et al. Piezoelectric PU/PVDF electrospun scaffolds for wound healing applications. Colloids and Surfaces B: Biointerfaces. 2012;**96**:29-36. DOI: 10.1016/j.colsurfb.2012.03.014

[36] Dong Z-Q, Ma X-h, Xu Z-L, You W-T, Li F-b. Superhydrophobic PVDF–PTFE electrospun nanofibrous membranes for desalination by vacuum membrane distillation. Desalination. 2014;**347**:175-183. DOI: 10.1016/j.desal.2014.05.015

[37] Choi S-S, Lee YS, Joo CW, Lee SG, Park JK, Han K-S. Electrospun PVDF nanofiber web as polymer electrolyte or separator. Electrochimica Acta. 2004;**50**:339-343. DOI: 10.1016/j.electacta.2004.03.057

[38] Gao K, Hu X, Dai C, Yi T. Crystal structures of electrospun PVDF membranes and its separator application for rechargeable lithium metal cells. Materials Science and Engineering: B. 2006;**131**:100-105. DOI: 10.1016/j.mseb.2006.03.035

[39] Liao Y, Wang R, Tian M, Qiu C, Fane AG. Fabrication of polyvinylidene fluoride (PVDF) nanofiber membranes by electro-spinning for direct contact membrane distillation. Journal of Membrane Science. 2013;**425**:30-39. DOI: 10.1016/j.memsci.2012.09.023

[40] Ding Y, Zhang P, Long Z, Jiang Y, Xu F, Di W. Preparation of PVdF-based electrospun membranes and their application as separators. Science and Technology of Advanced Materials. 2008;**9**:015005. DOI: 10.1088/1468-6996/9/1/015005

[41] Zhao Z, Zheng J, Wang M, Zhang H, Han CC. High performance ultrafiltration membrane based on modified chitosan coating and electrospun nanofibrous PVDF scaffolds. Journal of Membrane Science. 2012;**394**:209-217. DOI: 10.1016/j.memsci.2011.12.043

[42] Liu Z, Pan C, Lin L, Huang J, Ou Z. Direct-write PVDF nonwoven fiber fabric energy harvesters via the hollow cylindrical near-field electrospinning process. Smart Materials and Structures. 2013;**23**:025003. DOI: 10.1088/0964-1726/23/2/025003

[43] Yang Y, Centrone A, Chen L, Simeon F, Hatton TA, Rutledge GC. Highly porous electrospun polyvinylidene fluoride (PVDF)-based carbon fiber. Carbon. 2011;**49**:3395-3403. DOI: 10.1016/j.carbon.2011.04.015

[44] Liu X, Kuang X, Xu S, Wang X. High-sensitivity piezoresponse force microscopy studies of single polyvinylidene fluoride nanofibers. Materials Letters. 2017;**191**:189-192. DOI: 10.1016/j.matlet.2016.12.066

[45] Baji A, Mai Y-W, Li Q, Liu Y. Electrospinning induced ferroelectricity in poly (vinylidene fluoride) fibers. Nanoscale. 2011;**3**:3068-3071. DOI: 10.1039/C1NR10467E

[46] Liu Z, Pan C, Lin L, Lai H. Piezoelectric properties of PVDF/MWCNT nanofiber using near-field electrospinning. Sensors and Actuators A: Physical. 2013;**193**:13-24. DOI: 10.1016/j.sna.2013.01.007

[47] Salimi A, Yousefi A. Analysis method: FTIR studies of β-phase crystal formation in stretched PVDF films. Polymer Testing. 2003;**22**:699-704. DOI: 10.1016/S0142-9418(03)00003-5

[48] Liu X, Deng M, Wang X. Nanoscale domain imaging and local piezoelectric coefficient d 33 studies of single piezoelectric polymeric nanofibers. Materials Letters. 2017;**189**:66-69. DOI: 10.1016/j.matlet.2016.11.044

[49] Song Y, Shen Y, Liu H, Lin Y, Li M, Nan C-W. Enhanced dielectric and ferroelectric properties induced by dopamine-modified $BaTiO_3$ nanofibers in flexible poly (vinylidene fluoride-trifluoroethylene) nanocomposites. Journal of Materials Chemistry. 2012;**22**:8063-8068. DOI: 10.1039/C2JM30297G

[50] Guzhova A, Galikhanov M, Gorokhovatsky YA, Temnov D, Fomicheva E, Karulina E, et al. Improvement of polylactic acid electret properties by addition of fine barium titanate. Journal of Electrostatics. 2016;**79**:1-6. DOI: 10.1016/j.elstat.2015.11.002

[51] Huang Z-X, Liu X, Wu J, Wong S-C, Chase GG, Qu J-P. Electrospun poly (vinylidene fluoride) membranes functioning as static charge storage device with controlled crystalline phase by inclusions of nanoscale graphite platelets. Journal of Materials Science. 2018; **53**:3038-3048. DOI: 10.1007/s10853-017-1723-0

[52] Huang Z-X, Liu X, Zhang X, Wong S-C, Chase GG, Qu J-P, et al. Electrospun polyvinylidene fluoride containing nanoscale graphite platelets as electret membrane and its application in air filtration under extreme environment. Polymer. 2017;**131**:143-150. DOI: 10.1016/j.polymer.2017.10.033

[53] Lee JS, Kim GH, Hong SM, Choi HJ, Seo Y. Surface functionalization of a poly (vinylidene fluoride): Effect on the adhesive and piezoelectric properties. ACS Applied Materials & Interfaces. 2009;**1**:2902-2908. DOI: 10.1021/am900616s

[54] Kim SJ, Raut P, Jana SC, Chase G. Electrostatically active polymer hybrid aerogels for airborne nanoparticle filtration. ACS Applied Materials & Interfaces. 2017;**9**:6401-6410. DOI: 10.1021/acsami.6b14784

[55] Chattopadhyay S, Hatton TA, Rutledge GC. Aerosol filtration using electrospun cellulose acetate fibers. Journal of Materials Science. 2016;**51**:204-217. DOI: 10.1007/s10853-015-9286-4

[56] Yarin A, Kataphinan W, Reneker DH. Branching in electrospinning of nanofibers. Journal of Applied Physics. 2005;**98**:064501. DOI: 10.1063/1.2060928

[57] Kilic A, Shim E, Pourdeyhimi B. Effect of annealing on charging properties of electret fibers. The Journal of The Textile Institute. 2017;**108**:987-991. DOI: 10.1080/00405000.2016.1207269

[58] Pearce GK. UF/MF pre-treatment to RO in seawater and wastewater reuse applications: A comparison of energy costs. Desalination. 2008;**222**:66-73. DOI: 10.1016/j.desal.2007.05.029

[59] Kim SJ, Ko SH, Kang KH, Han J. Direct seawater desalination by ion concentration polarization. Nature Nanotechnology. 2010;**5**:297-301. DOI: 10.1038/nnano.2010.34

[60] Liao Y, Wang R, Fane AG. Engineering superhydrophobic surface on poly (vinylidene fluoride) nanofiber membranes for direct contact membrane distillation. Journal of Membrane Science. 2013;**440**:77-87. DOI: 10.1016/j.memsci.2013.04.006

[61] Alturkmani A. Industrial Wastewater. 2013

[62] Dermentzis K. Removal of nickel from electroplating rinse waters using electrostatic shielding electrodialysis/electrodeionization. Journal of Hazardous Materials. 2010;**173**: 647-652. DOI: 10.1016/j.jhazmat.2009.08.133

[63] Wu N, Wei H, Zhang L. Efficient removal of heavy metal ions with biopolymer template synthesized mesoporous titania beads of hundreds of micrometers size. Environmental Science & Technology. 2011;**46**:419-425. DOI: 10.1021/es202043u

[64] Kurniawan TA, Chan GYS, Lo W-H, Babel S. Physico–chemical treatment techniques for wastewater laden with heavy metals. Chemical Engineering Journal. 2006;**118**:83-98. DOI: 10.1016/j.cej.2006.01.015

[65] Nasreen SAAN, Sundarrajan S, Nizar SAS, Balamurugan R, Ramakrishna S. Advancement in electrospun nanofibrous membranes modification and their application in water treatment. Membranes. 2013;**3**:266-284. DOI: 10.3390/membranes3040266

[66] Tao Y, Ye L, Pan J, Wang Y, Tang B. Removal of Pb (II) from aqueous solution on chitosan/ TiO_2 hybrid film. Journal of Hazardous Materials. 2009;**161**:718-722. DOI: 10.1016/j.jhazmat.2008.04.012

[67] Zhou Y-T, Nie H-L, Branford-White C, He Z-Y, Zhu L-M. Removal of Cu^{2+} from aqueous solution by chitosan-coated magnetic nanoparticles modified with α-ketoglutaric acid. Journal of Colloid and Interface Science. 2009;**330**:29-37. DOI: 10.1016/j.jcis.2008.10.026

[68] Bottino A, Capannelli G, Comite A. Preparation and characterization of novel porous PVDF-ZrO_2 composite membranes. Desalination. 2002;**146**:35-40. DOI: 10.1016/S0011-9164(02)00469-1

[69] Liang S, Xiao K, Mo Y, Huang X. A novel ZnO nanoparticle blended polyvinylidene fluoride membrane for anti-irreversible fouling. Journal of Membrane Science. 2012;**394**: 184-192. DOI: 10.1016/j.memsci.2011.12.040

[70] Liu F, Abed MRM, Li K. Preparation and characterization of poly (vinylidene fluoride) (PVDF) based ultrafiltration membranes using nano γ-Al_2O_3. Journal of Membrane Science. 2011;**366**:97-103. DOI: 10.1016/j.memsci.2010.09.044

[71] Du J, Wu L, Tao C, Sun C. Preparation and characterization of Fe_3O_4/PVDF magnetic composite membrane. Acta Physico-Chimica Sinica. 2004;**20**:598-601. DOI: 10.3866/PKU.WHXB20040609

[72] Trigo CEL, Porto AO, De Lima GM. Characterization of CdS nanoparticles in solutions of P (TFE-co-PVDF-co-prop)/N, N-dimethylformamide. European Polymer Journal. 2004;**40**: 2465-2469. DOI: 10.1016/j.eurpolymj.2004.06.027

[73] Hashim NA, Liu Y, Li K. Preparation of PVDF hollow fiber membranes using SiO_2 particles: The effect of acid and alkali treatment on the membrane performances. Industrial & Engineering Chemistry Research. 2011;**50**:3035-3040. DOI: 10.1021/ie102012v

[74] Rahimpour A, Jahanshahi M, Rajaeian B, Rahimnejad M. TiO_2 entrapped nano-composite PVDF/SPES membranes: Preparation, characterization, antifouling and antibacterial properties. Desalination. 2011;**278**:343-353. DOI: 10.1016/j.desal.2011.05.049

[75] Zhang X, Wang Y, Liu Y, Xu J, Han Y, Xu X. Preparation, performances of PVDF/ZnO hybrid membranes and their applications in the removal of copper ions. Applied Surface Science. 2014;**316**:333-340. DOI: 10.1016/j.apsusc.2014.08.004

[76] Song L, Zhao X, Fu J, Wang X, Sheng Y, Liu X. DFT investigation of Ni (II) adsorption onto MA-DTPA/PVDF chelating membrane in the presence of coexistent cations and organic acids. Journal of Hazardous Materials. 2012;**199**:433-439. DOI: 10.1016/j.jhazmat.2011.11.046

[77] Salehi E, Madaeni SS, Heidary F. Dynamic adsorption of Ni (II) and cd (II) ions from water using 8-hydroxyquinoline ligand immobilized PVDF membrane: Isotherms, thermodynamics and kinetics. Separation and Purification Technology. 2012;**94**:1-8. DOI: 10.1016/j.seppur.2012.04.004

[78] WHO—Geneva C. Desalination for Safe Water Supply: Guidance for the Health and Environmental Aspects Applicable to Desalination. Available Online. Geneva, Switzerland: World Health Organization (WHO); 2007

[79] Kaur S, Sundarrajan S, Rana D, Matsuura T, Ramakrishna S. Influence of electrospun fiber size on the separation efficiency of thin film nanofiltration composite membrane. Journal of Membrane Science. 2012;**392**:101-111. DOI: 10.1016/j.memsci.2011.12.005

[80] Wang G, Pan C, Wang L, Dong Q, Yu C, Zhao Z, et al. Activated carbon nanofiber webs made by electrospinning for capacitive deionization. Electrochimica Acta. 2012;**69**:65-70. DOI: 10.1016/j.electacta.2012.02.066

[81] Prince JA, Singh G, Rana D, Matsuura T, Anbharasi V, Shanmugasundaram TS. Preparation and characterization of highly hydrophobic poly (vinylidene fluoride)–clay nanocomposite nanofiber membranes (PVDF–clay NNMs) for desalination using direct contact membrane distillation. Journal of Membrane Science. 2012;**397**:80-86. DOI: 10.1016/j.memsci.2012.01.012

[82] Hoover LA, Schiffman JD, Elimelech M. Nanofibers in thin-film composite membrane support layers: Enabling expanded application of forward and pressure retarded osmosis. Desalination. 2013;**308**:73-81. DOI: 10.1016/j.desal.2012.07.019

[83] Lalia BS, Guillen-Burrieza E, Arafat HA, Hashaikeh R. Fabrication and characterization of polyvinylidenefluoride-co-hexafluoropropylene (PVDF-HFP) electrospun membranes for direct contact membrane distillation. Journal of Membrane Science. 2013;**428**:104-115. DOI: 10.1016/j.memsci.2012.10.061

[84] Mei Y, Yao C, Fan K, Li X. Surface modification of polyacrylonitrile nanofibrous membranes with superior antibacterial and easy-cleaning properties through hydrophilic flexible spacers. Journal of Membrane Science. 2012;**417**:20-27. DOI: 10.1016/j.memsci.2012.06.021

[85] Tijing LD, Woo YC, Johir MAH, Choi J-S, Shon HK. A novel dual-layer bicomponent electrospun nanofibrous membrane for desalination by direct contact membrane distillation. Chemical Engineering Journal. 2014;**256**:155-159. DOI: 10.1016/j.cej.2014.06.076

[86] RE BUXBAUM. Membrane reactor advantages for methanol reforming and similar reactions. Separation Science and Technology. 1999;**34**:2113-2123. DOI: 10.1081/SS-100100759

[87] Cardea S, Reverchon E. Nanostructured PVDF-HFP membranes loaded with catalyst obtained by supercritical CO_2 assisted techniques. Chemical Engineering and Processing: Process Intensification. 2011;**50**:630-636. DOI: 10.1016/j.cep.2011.03.006

[88] Bottino A, Capannelli G, Comite A, Di Felice R. Polymeric and ceramic membranes in three-phase catalytic membrane reactors for the hydrogenation of methylenecyclohexane. Desalination. 2002;**144**:411-416. DOI: 10.1016/S0011-9164(02)00352-1

[89] Samanta S, Nandi AK. Recyclable high performance palladium heterogeneous catalyst with porous PVDF binder from the novel gel nanocomposite route. The Journal of Physical Chemistry C. 2009;**113**:4721-4725. DOI: 10.1021/jp810160r

[90] Gotardo MC, Guedes AA, Schiavon MA, José NM, Yoshida IVP, Assis MD. Polymeric membranes: The role this support plays in the reactivity of the different generations of metalloporphyrins. Journal of Molecular Catalysis A: Chemical. 2005;**229**:137-143. DOI: 10.1016/j.molcata.2004.11.014

[91] Shore SG, Ding E, Park C, Keane MA. Vapor phase hydrogenation of phenol over silica supported Pd and Pd.Yb catalysts. Catalysis Communications. 2002;**3**:77-84. DOI: 10.1016/S1566-7367(02)00052-3

Mathematical Modeling of the Relation between Electrospun Nanofibers Characteristics and the Process Parameters

Liliana Rozemarie Manea, Andrei-Petru Bertea,
Elena Nechita and Carmen Violeta Popescu

Additional information is available at the end of the chapter

http://dx.doi.org/10.5772/intechopen.75350

Abstract

Electrospinning, the most favorable process of obtaining nanofibers, is capable of processing solution or melt polymers, ceramic materials or metals in many morphological variants, thus providing diverse functionalities. The chapter reviews the main ways in which nanofibers' characteristics can be influenced by solution parameters, process parameters and ambient conditions, afterwards focusing on the role of some of the most significant electrospinning parameters (applied voltage, flow rate, nozzle to collector distance) on the diameter of the nanofibers. Experimental studies to model the influence of process parameters in the case of electrospinning polyetherimide solutions are presented. Response surface methodology and MATLAB simulation software have been used to obtain the mathematical models that indicate the most favorable parameters.

Keywords: electrospinning, polyetherimide, process parameters, nanofiber diameter, mathematical model

1. Introduction

Electrospinning is an extremely flexible technique, being applicable to both polymer solutions and melts, which can be converted into nanometer-grade fiber. This method offers access to completely new materials, which may have complex chemical structures and a very wide range of usage areas.

Electrospun nanofibers have many applications in filtration processes, bio-medicine (tissue engineering, drug delivery, scaffolds and wound healing), in energy devices and sensors, depending on their morphological characteristics. The morphology of the fibers obtained through electrospinning is governed by many factors. These factors are related to the polymer solution characteristics, process and ambient parameters.

2. Main electrospinning parameters affecting nanofibers morphology

The morphology of the nanofibers obtained through electrospinning is influenced by many factors. These factors can be grouped into three categories: polymer solution characteristics, process parameters and ambient parameters. Knowing the way these factors influence the electrospinning process, it becomes easier to obtain nanofibers with controlled structure and required functions [1, 2].

2.1. Polymer solution

2.1.1. Viscosity

Some of the key parameters of the electrospinning process are the surface tension and the viscoelastic properties of the polymer solution [3–5]. The correct choice of the concentration of the polymer solution has a decisive effect on these processes, respectively, on the characteristics of the obtained nanofibers (diameter and morphology) [6, 7].

It is generally accepted that the viscosity of the polymer solution is the decisive parameter of the process, including the possibility of electrospinning and the characteristics of the nanofibers. In order to make electrospinning possible, the viscosity must be in a relatively narrow range. At very low viscosity values, the polymer filaments break and polymer droplets are produced, while at very high viscosity values, the polymer solution cannot pass through the nozzles and electrospinning does not take place. The area of optimum viscosity depends essentially on three parameters: the nature of the polymer, the nature of the solvent and the concentration of the polymer solution.

Figure 1 shows the great significance of the polymer solution viscosity for electrospinning.

Usually, a solution having a viscosity of 1–20 poise and whose surface tension is in the range of 35–55 dyn/cm^2 is considered to be suitable for electrospinning. If the viscosity values exceed the upper limit of the abovementioned range, the cohesion of the solution greatly increases in order to achieve stable polymer flow. Large beds are formed, with great distance between them, which generates larger fiber diameters. For polymeric solutions with a viscosity of less than 1 poise, the breaking of the polymer flow is recorded, with the formation of frequent beads [9]. With the increase in the solution concentration, there is a change in the shape of the beds, which passes from spherical to spindle-like [10].

There is a direct proportionality relationship between the concentration of the electrospun polymer solution and the diameter of the obtained nanofibers, but the value of the ratio of proportionality varies within relatively large limits.

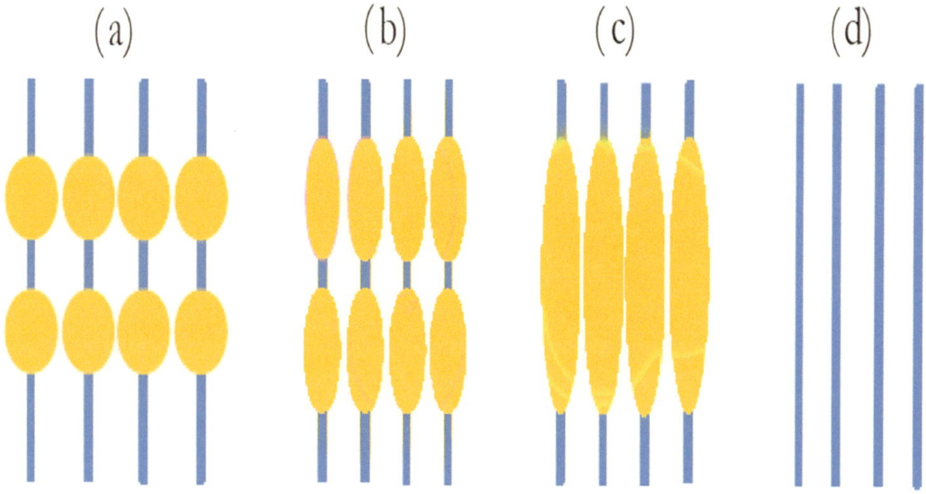

Figure 1. The viscosity increases from (a) to (d): (a) drops; (b) slightly extended drops; (c) extended drops; (d) nanofibers.

However, the use of decreasing the concentration of the polymer solution to reduce the diameter of the fibers is limited by inherent difficulties in electrospinning or even the impossibility of electrospinning, as well as by problems in diminishing the diameter by subsequent treatments [11]. It is worth mentioning the possibility of increasing the surface of the fibers by further modification of their morphology either by changing the shape of their cross-section or by forming pores, pits or bumps on their surface [12].

The close relationship between the viscosity and the concentration of the polymer solution to be electrospun has been studied for polymers such as poly (lactic-co-glycolic acid) (PLGA) [13], poly (ethylene oxide) (PEO) [14, 15], poly (vinyl alcohol) (PVA) [16–18], poly (methyl methacrylate) (PMMA) [19], polystyrene (PS) [20], poly (L-lactic acid) (PLLA) [8], gelatin [21] and dextran [22]). A study on the relationship between viscosity and concentration on electrospinning of seven solutions of linear homopolymers of poly (methyl methacrylate) with diverse molecular weights (dimethylformamide—DMF—was the solvent) demonstrated that when electrospinning solutions with concentrations below the value obtained by multiplying the overlap concentration by six (the overlap concentration is the concentration of the polymer solution for which there is a sudden change in viscosity) no filaments are obtained, but only beads, while high molecular weight PMMA at concentrations in this range produces fibers having a relatively limited number of beds. [18].

A new method of reducing the viscosity of the polymer solution is the application of low-frequency vibrations during electrospinning, when the polymer chains are untangled as a result of the breakage of the interchain van der Waals bonds. In the case of a PMMA solution to which a vibration of 300 rad/sec was applied, the viscosity diminution was achieved by an order of magnitude [23, 24]. The application of vibrations with a frequency of about 400 Hz on the tip of the capillary when electrospinning a solution of poly (butylene succinate) (PBS)/$CHCl_3$ led to a significant reduction in the diameter of the obtained nanofibers [25]. The use of vibrations may make possible the electrospinning of more concentrated polymer solutions that would otherwise not be electrospun, as well as suspensions or coagulated materials [12].

2.1.2. Conductivity

Since electrospinning is essentially based on the process of passing electrical charges from the electrode to the polymer solution at the tip of the injection needle, it is imperative that the polymer solution exhibits a certain level of electrical conductivity without which the process cannot be realized.

The conductivity of the polymer solution means the number of electrical charges found on the surface of the solution, the presence of which determines the formation of nanofibers rather than nanoparticles.

Upon dissolution of a polymer, an increase in the conductivity of the solution is recorded because the various polymeric ionic species (mostly impurities or additives) are available in this way. It has been found that a decrease in conductivity can be observed if increasing the concentration of the polymer solution [26]. In the particular case when the polymer is a polyelectrolyte, its solution will exhibit high conductivity, its value depending largely on the concentration of the solution [21].

The addition of electrolytes leads to an increased number of electrical charges, thus improving the conductivity, which leads to the increase of the elongation capacity of the polymer solution, the fibers obtained in this case being smoother and finer [13].

For conductivity enhancement, inorganic salts such as NaCl (0.01 M) [6, 14, 15] or ionic organic compounds such as pyridinium formate [26], palladium diacetate [16], chloride trialkylbenzyl ammonium can be added to the solution [17, 18].

The morphology of the nanofibers thus obtained is influenced by the dimensions of the ions introduced into the electrospinning solution, there being a relationship of inverse proportionality between the ion size and the uniformity of the nanofibers. In addition, for small dimensions of the added ions, the number of defects decreases [19].

2.1.3. Surface tension

The surface tension is a property of liquids that makes them take a geometric shape of minimum area in the absence of external forces, due to the cohesion forces between molecules [20].

The surface tension has an important significance in the electrospinning process. For example, lower surface tension will allow electrospinning to be achieved at an inferior electric field [28, 30–34]. Because it depends on the characteristics of the solvent, its correct choice is of the utmost importance, both for a homogeneous solution and for a consistent surface tension.

As a rule, it is considered that, for a given situation, surface tension determines the limits of the range in which electrospinning can be achieved [23–25, 27–29].

When an electric field is not applied, the surface tension of the solution causes the solution to be retained at the tip of the capillary. When applying an electric field, the electric charges exceed the forces of superficial tension and the polymer jet forms. Decreasing the surface tension of the polymer solution can cause the instability of the polymer jet, whose tendency to break and form droplets increases [1, 35–39].

Proper adjustment of the surface tension can be achieved by choosing the correct polymer solution concentration [40–44] or by adding a surfactant; in this second case, the uniformity of the electrospun fibers increases. However, the mere presence of a low superficial tension of a polymer solution does not guarantee the possibility of electrospinning it [20, 45].

2.2. Process parameters

Process variables, such as fluid flow rate, spinneret to collector distance and electric field strength have a significant influence on the properties of the electrospun nanofibers.

2.2.1. Fluid flow rate

Many of the elements of the electrospinning process, such as the initial shape of the droplet, the persistence of the Taylor cone, the trajectory of the extruded jet, the area in which the nanofibers are deposited and their essential characteristics, such as the size and uniformity of the diameter, are controlled by the flow rate of the polymer solution into the syringe.

As a rule, lower flow rate allows longer polarization times; if this speed exceeds a certain threshold value, a value that depends on the nature of the polymer and the solvent, the feed rate of the polymer solution exceeds the rate at which it is pulled from the tip due to the applied electrical forces. In this situation, since the solidification time of the polymer filament until it reaches the collector is too low, the obtained fibers have a high density of beads and an increased number of large droplets.

The way the shape of the polymer cone changes with increasing flow rate is shown in **Figure 2**.

When electrospinning a polystyrene solution, it was found that the formation of beads occurs when the flow rate exceeds 0.1 mL/min, conditions in which fiber diameter and pore size increase [47]. When electrospinning a solution of 20% polysulfone in N, N-dimethylacetamide at 10 kV, it was observed that at a feed rate of 0.66 mL/h, smaller diameters of fiber were obtained [48], while when electrospinning a nylon 6 solution, bead fiber formation occurred when the flow rate surpassed 4 mL/min [49].

The flow rate can be increased, in association with increased applied stress, to improve the productivity of the process, obtaining thinner nanofibers. If the flow rate is very much increased webs may be obtained instead of fibers, as the solution ejected from the tip does not have enough time to dry until reaching the collector [46].

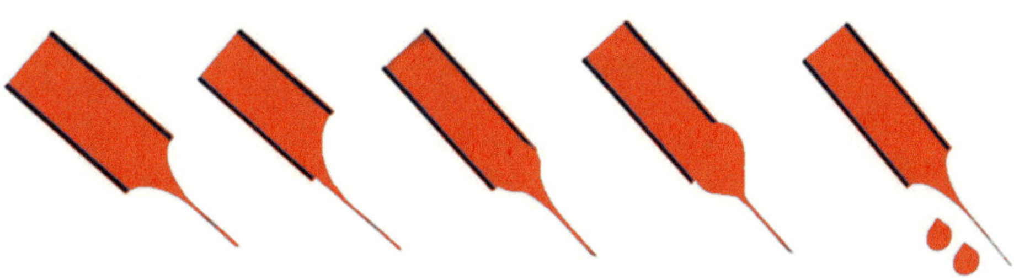

Figure 2. Development of various jets when flow rate increases.

In addition to this, if the flow rate is increased, a broader fiber diameter distribution is obtained. Because too high or too low flow rates influence the electrospinning process and the diameter of the resulted nanofibers, it is better to keep a flow rate as low as to ensure the equilibrium between the extruded polymer solution and the fresh replacing solution during the formation of the jet [2].

Choosing an appropriate flow rate leads to a limitation of the formation of defects such as blobs, splitting and branched fibers. In addition, a stable and constant flow rate is required [50].

2.2.2. Distance between the spinneret and the collector

The morphology and the diameter of the electrospun fibers can be controlled by proper setting of the distance between the spinneret and the collector.

If this distance is too small, the fibers will not solidify before reaching the collector, while if the distance is too long, droplets may appear on the surface of the fibers [1, 21]. The distance between the syringe tip and the collector has to be judiciously set, as the speed of the solvent removal is very important in order to produce quality nanofibers. Even the smallest modification of this distance can have serious effects on fiber characteristics.

Additionally, if the distance between the nozzle and the collector is increased, the level of the electric field between the two decreases, forming fewer charged ions, boosting bending instability and elongation tendency and decreasing the diameter of the polymer jet [50, 51].

Many papers show that a small spinneret to collector distance leads to faulty and large-diameter nanofibers, while the nanofibers diameter decreases as the distance is increased [52]. In the case of poly (vinyledene fluoride) nanofibers obtained through electrospinning from a 28 wt% solution at 12 kV, a decrease in diameter from 397 nm to 314 nm was observed when the distance between the nozzle and the collector increased from 15 cm to 16 cm, associated with better uniformity [53].

A study on the influence of the distance between the spinneret and the collector on the diameter of nanofibers electrospun from 12% polyetherimide solution, obtained using a mixture of dimethylacetamide /tetrahydrofuran 1:1 as solvent, led to the conclusion that the smallest values of the diameter are obtained for a distance of 45 mm [44].

Usually the appropriate distance between the spinneret and the collector differs from one polymer system to another, but there were cases when no effect on the morphology of the nanofiber occurred when altering the distance between the needle and collector [54].

2.2.3. Applied voltage

The applied electrical voltage is considered to be one of the most significant parameters of the electrospinning process, as it drastically affects both the dynamics of the fluid flow and the morphology of the electrospun fibers. Its impact depends upon the concentration of the polymer solution and the distance between the spinneret and the collector [55].

The applied voltage is important because the charged polymer jets leave the Taylor cone only if the applied voltage exceeds a specific threshold value, which depends on both the type of polymer and the type of solvent [2, 38].

The way the applied stress influences the morphology of the electrospun nanofibers is subject to controversy.

Some studies identify a relationship of inverse proportionality between the applied voltage and the diameter of the electrospun nanofibers. The argument for this behavior is the more prolonged elongation of the extruded polymer filament when the rejection forces are greater [56, 57]. There are also studies that did not find a correlation between the two parameters when electrospinning a poly (ethylene oxide) solution [49, 58]. Moreover, in case of aqueous poly (vinyl alcohol) solution electrospinning, the increase in the applied voltage leads to an increase in the diameter of the nanofibers [54]. It can be concluded that the applied voltage value frequently affects the diameter of the obtained nanofibers, but the type of influence depends on the nature of the polymer / solvent system and the distance between the electrode and the collector [59].

Expert opinions are far less divergent as to the relationship between the value of the applied voltage and the probability of defect formation. It has been found that an increase in the applied tension leads to an increase in the deposition rate, which explains the augmentation of the number of defects [56, 60]. The length of electrospun nanofibers decreases when the applied voltage increases, without affecting the pore size [34, 61]. It was found that in most cases a lower applied voltage will cause the production of nanofibers with uniform morphology and with low number of defects [62].

The study on a mixture of polyaniline-camphor sulfonic acid/poly (ethylene oxide) highlighted the fact that higher voltage leads to thinner nanofibers, but also a greater diversity of diameters and a wider distribution of diameters [64].

2.3. Ambient conditions

Any interaction between the environment and the polymer solution may affect the electrospun fibers' morphology, with the greatest impact given by humidity, temperature and atmospheric pressure.

2.3.1. Humidity

The role of the humidity of the electrospinning environment is manifested in terms of fiber morphology, deposition orientation and solvent evaporation rate. When the humidity is very low, a volatile solvent can dry very quickly. It also has been suggested that high humidity helps discharging static electricity from electrospun fibers.

At high humidity, condensation may occur at the fiber surface due to the cooling of the jet surface, caused by the rapid evaporation of the volatile solvent, and the air flow may interrupt the formation of the fibers, causing fiber breaking [34].

An experiment on polysulfide electrospun nanofibers [63] revealed that at a moisture content of less than 25%, the fibers are smooth, glossy, with non-porous surface; at a humidity of

31–38% there is a visible difference in fiber morphology, with a relatively small number of randomly distributed circular pores; at 31–45% humidity the pores move toward the surface of the fiber; at a humidity of 50–59% the pores are numerous at the fiber surface; while at a moisture content of 60–72% the pores are larger and more uneven because of the incomplete drying of the electrospun fibers and their entanglement on the surface of the collector.

When electrospinning a polyaniline-camphor sulfonic acid/poly (ethylene oxide) mixture, it was found that only low ambient humidity allowed the formation of defect-free nanofibers over the entire range of applicable voltages. When the humidity reaches 25%, fibers start to break forming uneven and irregular surfaces, while at relative humidity higher than 40%, electrospinning is no longer possible [64].

2.3.2. Temperature

The temperature of the electrospinning ambient significantly influences the process. A first effect is manifested on the evaporation rate of the solvent, which exponentially decreases with decreasing temperature; the evaporation process of the solvent becomes slower, the jet takes a longer time to solidify, which can lead to defects in fiber formation [35].

Temperature has an important influence on the rigidity of the polymeric chains, which decreases as the ambient temperature increases. Under these conditions, associated with a low viscosity of the polymer solution, it is possible to obtain a better stretch of the polymer filament under the action of Coulombic forces, resulting in smaller diameter fibers.

2.3.3. Atmospheric pressure

At low atmospheric pressure, the polymer solution in the syringe tends to flow, causing unstable jet initiation, and at very low pressures electrospinning cannot take place due to direct discharge of electrical charges.

3. Mathematical models of the electrospinning of polyetherimide

Our study investigated the way the distance between electrodes and some of the main technological parameters affect the diameter of the electrospun polyetherimide fibers [39, 44]. A 12% polyetherimide solution, obtained using as solvent a mixture of dimethylacetamide / tetrahydrofuran (1:1 ratio), has been used. The spinning equipment, which worked at normal atmosphere, is of multijet type, with needle and uniaxial delivery.

In earlier experiments [43, 65, 66], the solubility of polyetherimide in the dimethylacetamide/tetrahydrofuran solvent (DMAC/THF) has been tested by computing the Hansen coefficients, and the results proved that the polymer has very good solubility at 12% concentration, unlike other tested concentrations (8–14%).

The properties of the 12% polyetherimide solution in the DMAC/THF mix [44, 67] are conductivity 1.18 mS/cm, surface tension 30.3 mN/m, zero shear viscosity 0.191 Pas.

In order to obtain the polymer solution, the polymer was dried for 2 hours at 100°C in vacuum. The polymer dissolution in the solvent mix was made by magnetic stirring for 24 h at 500°C.

The experimental equipment that has been used has three 3 mL syringes with 0.2 mm inner needle diameter and an inter-nozzle distance of 2.5 mm and a rotating cylinder type collecting mechanism, with cylinder rotation speed v = 1000 rpm. The displacement range along the Ox axis was 100 and 80 mm along the Oz axis.

The parameters that have been varied in the experiment were the distance between the needles and the collector (45, 70, 100, 120, 120 mm) and the technological parameters: flow rate (0.05, 0.075, 0.1, 0.15 mL/min) and the voltage (15, 20, 25, 30 and 35 kV).

Scanning electron microscopy (SEM) was used to characterize the obtained polyetherimide electrospun nanofibers; the nanofibers have been previously gold plated using a Phenom G2 equipment [39, 66]. To determine the diameter of the electrospun fibers, a Lucia image analysis software was used. For each technological variant, 100 measurements of the diameter of the electrospun fiber have been carried out [44, 66].

To analyze the experimental data, the Response Surface Methodology (RSM) was used as a mathematical and statistical technique. RSM was used to find the dependence between d_{med}, the mean fibers diameter, and D, the distance between needles and collector, in conjunction with the flow rate Q and the voltage U. The notations for the variable parameters are x1 for D, x2 for Q and x3 for U. These are the predictors of the model, which were varied together in accordance with the experimental plan, in order to determine the most favorable combinations of Q and U which gives the desired fiber diameters resulting from the electrospinning process.

The experiments were performed under the following environmental conditions: 20°C, RH = 40%, normal atmospheric pressure. Under these specified conditions, the values for spinning distance, applied voltage and volume flow rate have been selected to determine the influence of these parameters on the polyetherimide nanofiber electrospinning process.

In order to study the dependence between the average fibers diameter d_{med} and the distance between the needles and the collector, (D) correlated with the flow rate (Q) and the applied voltage (U), response surface methodology (RSM) has been used. All the three explanatory variables were varied at the same time, in compliance with the experimental design.

The relationship between the response variable d_{med} and the explanatory variables can be written as d_{med} = f(x1, x2, x3), where the function f is to be found. The multiple regression method has been applied to approximate f with a second order polynomial P(x1, x2, x3), but after applying the statistical tests it was found that the complexity of the model does not require such a high-accuracy approach. In this case, the research plan was focused on using RSM to investigate three partial dependencies: a) d_{med} = u(x1, x2), for all the five values of x3 = U considered in the experiment, b) d_{med} = v(x2, x3), for three values of x1 = D which are expected to be favorable for the aim of obtaining very small fiber diameters, according to the results obtained at (a) and (c) d_{med} = w(x1, x3), for two values of x2 = Q.

In addition to the generation of the quadratic models for u, v and w, RSM supplies three-dimensional charts of the response surfaces, which are useful in the visualization of the relationship between the response and the predictors (independent variables) and allow us to observe the influence of each predictor.

The experiment was conducted for five levels of D, four levels of Q and five levels of U. Therefore, there are 5 × 4 × 5 = 100 treatment combinations of parameters. **Table 1** presents the regression equations for $d_{med} = u(x1, x2)$ for each of the five considered values of U.

The adjusted R-squared has been used, as this statistic is the most significant in selecting the right regression model. This value is designed to avoid the problem with the regular R-squared value. When a new term is added in the model, R-squared increases, which can be misleading and can lead to choosing a too complex model, which is not necessarily the right choice. Usually, it has been chosen from the models with greater adjusted R-squared values, which means that a high fraction of data is fitted by the model. In contrast to the regular R-squared, the adjusted R-squared increases only if the new term improves the model more than would be expected under random conditions. The adjusted R^2 always has a lower value than regular R^2.

Figure 3(a)–(e) shows the diagrams of the response surfaces for $d_{med} = u(x1, x2)$, for the five values of U. The charts display the correlation between D and Q and their influence on d_{med}.

Moreover, to give an overview on the behavior of d_{med} for the five values of the parameter x3 = U, we have represented the five response surfaces within the same coordinates system in **Figure 3(f)**.

It appears that d_{med} increases as both x1 = D and x2 = Q increase. It is obvious that higher values of U (U = 25 kV, U = 30 kV and U = 35 kV) are the most favorable.

We have further determined the approximation models for the dependency $d_{med} = v(x2, x3)$, for D = 45 mm, D = 70 mm and D = 100 mm. The results are shown in **Figure 4**, where the chart displays a simultaneous representation of the three models. The charts indicate that d_{med} increases as D increases.

Finally, the approximation models for the dependency $d_{med} = w(x1, x3)$ have been determined. A chart with all the response surfaces, for the five values of Q, is given in **Figure 5**.

The statistical analysis has been performed using Matlab 7.5.0 (R2007b) and the technical interpretation of the results provided by the data analysis is given further.

No.	U (kV)	Regression equation	R^2_{adj}
(1)	15	$d_{med} = 356.7 - 1.55D + 5553.6Q + 14.59DQ - 0.02\,D^2 - 18834.5Q^2$	94.42%
(2)	20	$d_{med} = 480.1 + 0.44D - 522.4Q - 25.8DQ + 0.0055\,D^2 - 5535.3Q^2$	97.03%
(3)	24	$d_{med} = 411.6 - 0.81D + 348.7Q - 30.66DQ + 0.0087\,D^2 - 2018.1Q^2$	93.72%
(4)	30	$d_{med} = 349.7 - 1.2D + 186.3Q + 25.77DQ + 0.0125\,D^2 - 206.9Q^2$	93.34%
(5)	35	$d_{med} = 283.2 - 1.05D + 2759.6Q + 21.3DQ + 0.01\,8D^2 - 17259.6Q^2$	89.72%

Table 1. Mathematical models for mean fibers diameter as function of D and Q.

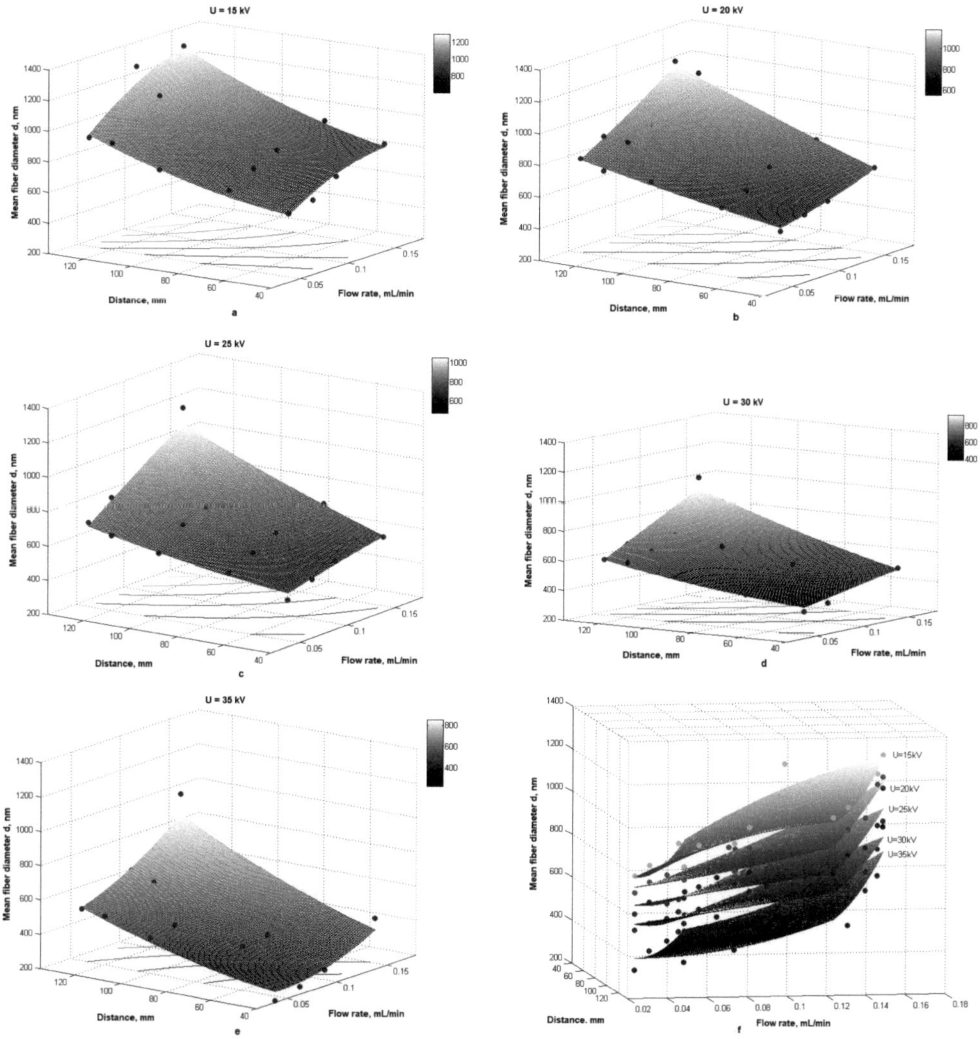

Figure 3. Response surfaces for mean fiber diameter in terms of spinning distance (D) and flow rate (Q) for the five values of U: (a) U = 15 kV, (b) U = 20 kV, (c) U = 25 kV, (d) U = 30 kV, (e) U = 35 kV, (f) Simultaneous representation for the five response surfaces.

As the presented charts display, longer spinning distance induces an increase in the fiber diameter. In Eqs. (1), (3)–(5) from **Table 1**, the coefficient of D is negative. In Eq. (2), the coefficient is positive, but very small. The literature reports both increase in fiber diameter and decrease in fiber diameter [45, 60, 65, 66] depending on D, due to different ratio between D and the electric field strength E (KV/cm), polymer solution concentration and solvent evaporation rate. In order to observe the influence of each term, the test for the individual coefficients was performed for the five models. The results of the stepwise regression show that the significant terms are those related to Q, D2 and Q2 for Eq. (1) and those related to DQ and D2 in Eqs. (2)–(5).

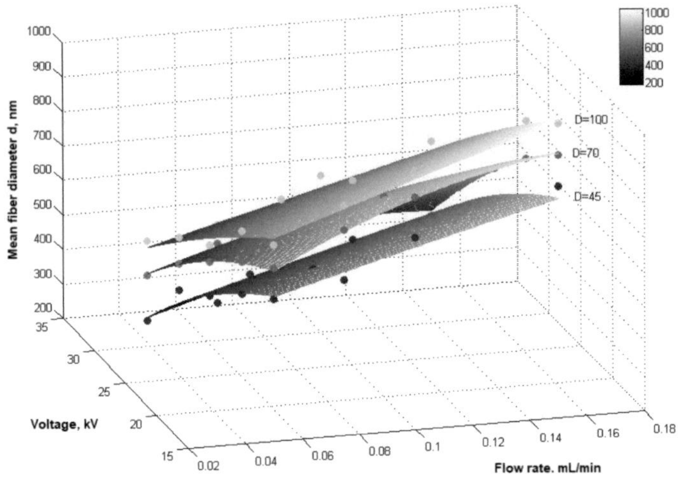

Figure 4. Response surfaces for mean fiber diameter in terms of voltage (U) and flow rate (Q) for three values of D. Simultaneous representation.

It can be concluded that the mathematical models as well as the observations on the response surfaces plotted for the partial dependencies (d_{med} as function of D (mm) and Q (mL/min), Q (mL/min) and U (kV), and as function of D (mm) and U (kV)) show that the optimum technological domain is defined by small values of the spinning distance (D = 45—70 mm). As we previously concluded, appropriate values of the mean fiber diameters are obtained for the smallest values of D (mm). The minimum value of fiber diameter is obtained for D = 45 mm, Q = 0.05 mL/min and U = 35 kV.

Next, our attention was focused on the results provided by RSM for d_{med} = v(x2, x3), for three values of x1 = D which are the most favorable for obtaining very small diameter fibers and for d_{med} = w(x1, x3), for two values of x2 = Q.

The model u(x1, x2) revealed that d_{med} decreases as x1 decreases. Therefore, in representing the response surfaces for the dependency d_{med} = v(x2, x3), we have considered for x1 = D only the three smallest values: D = 45 mm, D = 70 mm and D = 100 mm. The surfaces are represented in **Figure 6(a)–(c)**.

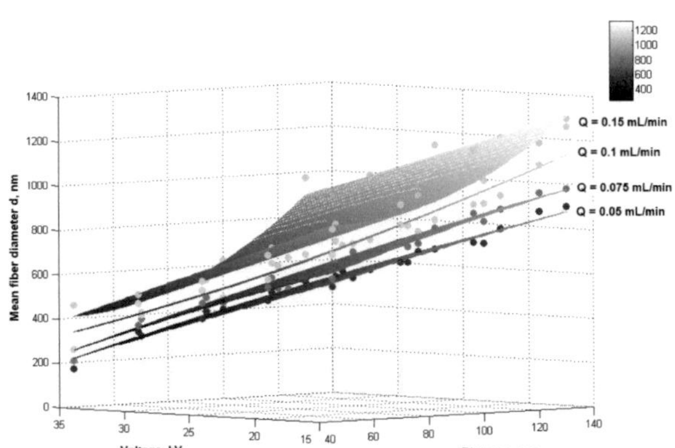

Figure 5. Response surfaces for mean fiber diameter in terms of spinning distance (D) and voltage (U). Simultaneous representation for the four values of Q.

http://dx.doi.org/10.5772/intechopen.75350

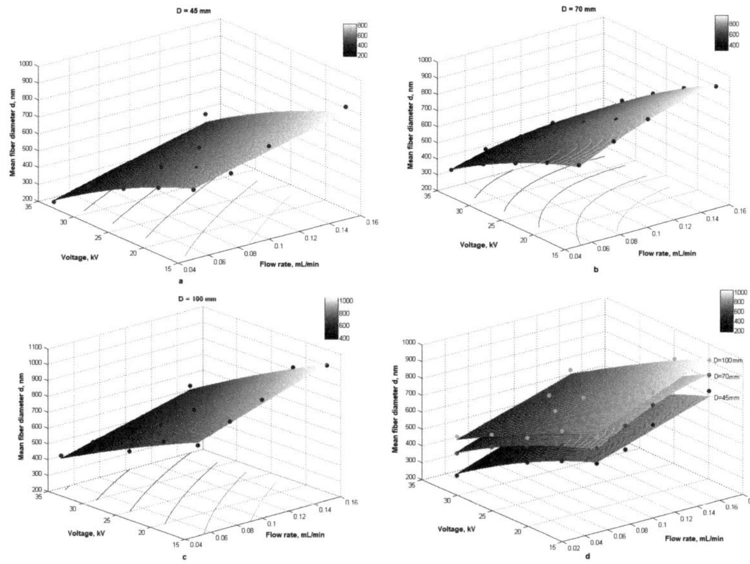

Figure 6. Response surfaces for mean fiber diameter in terms of voltage (U) and flow rate (Q) for three values of D: (a) D = 45 mm, (b) D = 70 mm, (c) D = 100 mm, (d) simultaneous representation for the three response surfaces.

It appears that small values of the flow rate entail small mean fiber diameters: d_{med} decreases as Q decreases. Therefore, the approximation models for the third partial dependency d_{med} = w(x1, x3) has been analyzed only for Q = 0.05 mL/min and Q = 0.075 mL/min (the lowest values of the flow rate). The graphic representation of this dependences (the response surfaces) are shown in **Figure 2(a)–(c)** and demonstrate that d_{med} decreases as x3 = U increases. In addition, it can be seen that d_{med} also increases with x1 = D as demonstrated previously by the model d_{med} = u(x1, x2).

Further the three models u(x1, x2), v(x2, x3), w(x1, x3) have been integrated, providing the technical interpretation of the results. This interpretation considers the behaviors and facts communicated on this topic, derived by other researchers [39, 55, 66] from similar experiments.

The response surfaces previously shown allow the visualization of the relationship between the response variables (the flow rate Q and voltage U) and the explanatory variables (the average fibers diameter in in correlation with the distance between the needles and the collector. The influence of the nozzle to collector distance D (symbolized by x1) on the fibers diameter has been previously analyzed. Numerous scientific articles dealing with this topic state that the fiber diameter may sometimes increase, but in other situations it decreases [39, 53, 55, 66] upon D. The main factors that influence this dependency are the ratio between nozzle to collector distance D and the electric field strength E, the concentration of the polymer solution and solvent evaporation rate. When electrospinning a polyetherimide solution, a mathematical model was found indicating that longer nozzle to collector distance induces an increase in the fibers diameter.

With respect to the influence of the flow rate Q = x2 (mL/min) on the mean fibers diameter, the experiments showed that d_{med} increased with the volume flow rate, as depicted in **Figure 3(a)–(f)** and also in **Figure 6(a)–(c)**. The Eqs. (3)–(5) in **Table 1** support this conclusion, as the coefficient of DQ is positive and the corresponding terms are significant. Our findings are consistent with previous research [28, 65].

Studying the influence of the voltage U = x3 (KV) on the mean fibers diameter, it was found that the diameter decreases when the voltage increases. The chart (a)–(c) in **Figure 6** and (a)–(b) in **Figure 7** also display this behavior. It can be observed that the minimum value of the fibers diameter is obtained for U = 35 kV, meaning that experimental data comply with the previous conclusion.

Figure 8 presents the projection of the response surface in **Figure 3(e)**, for the most favorable value of U, which leads to small fibers diameter, namely 35 kV. The dark-gray lines are displayed in the favorable zone of the parameters (while the light-gray ones are higher).

This study proved that the optimal technological domain is determined by small values of the nozzle to collector distance (D = 45–70 mm), small flow rate values (Q = 0.05–0.075 mL/min) and high applied voltage values (U = 30–35 kV). The desired characteristics for the fibers diameter are obtained when D (mm) and Q (mL/min) are minimal and U (kV) is maximum.

It can be concluded that the ability to control the diameter of electrospun nanofibers is of utmost importance as it affects the majority of the properties of the final product. The diameter of the electrospun nanofibers is dependent on a series of parameters, and in order to correctly assess the role of each parameter, the one-variable-at-a-time technique is not very eloquent. That is why addressing the problem by using experimental design can provide a more accurate picture of these dependences.

Based on our studies and many other scientific approaches in this area of research, it can be stated that an accurate choice of environmental and technological parameters for each

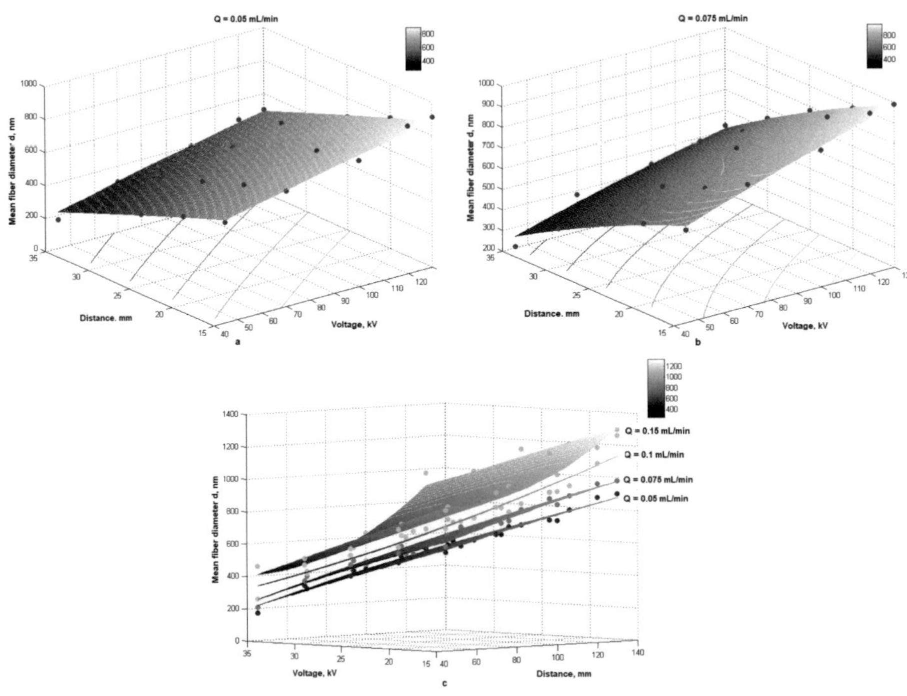

Figure 7. Response surfaces for mean fiber diameter in terms of distance (D) and voltage (U) for two values of Q: (a) Q = 0.05 mL/min, (b) Q = 0.075 mL/min, (c) simultaneous representation for all the values of Q.

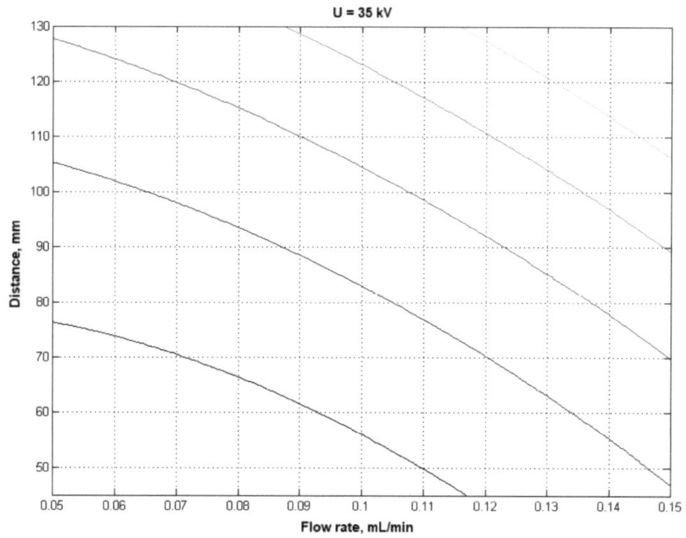

Figure 8. The projection of the response surface for U = 35 kV.

polymer solution (with optimal values of concentration, viscosity, molecular weight, solution conductivity) and a proper correlation of these parameters lead to obtaining flawless fibers with predetermined diameter.

Acknowledgements

This work was supported by a grant from the Romanian National Authority for Scientific Research and Innovation.

Author details

Liliana Rozemarie Manea[1], Andrei-Petru Bertea[1]*, Elena Nechita[2] and Carmen Violeta Popescu[2]

*Address all correspondence to: andrei_bertea@yahoo.co.uk

1 "Gheorghe Asachi" Technical University, Iași, Romania

2 "Vasile Alecsandri" University, Bacău, Romania

References

[1] Afshari M, editor. Electrospun Nanofibers. 1st ed. Cambridge, MA, USA: Woodhead Publishing; 2017. 648 p. DOI: 10.1016/c2014-0-04496-6

[2] Haider A, Haider S, Kang I-K. A comprehensive review summarizing the effect of electrospinning parameters and potential applications of nanofibers in biomedical and biotechnology. Arabian Journal of Chemistry. 2015;**8**(7):894-921. DOI: 10.1016/j.arabjc.2015.11.015

[3] Attout A, Yunus S, Bertrand P. Electrospinning and alignment of polyaniline-based nanowires and nanotubes. Polymer Engineering and Science. 2008;**48**(9):1661-1666. DOI: 10.1002/pen.20969

[4] Ginestra P, Ceretti E, Fiorentino A. Electrospinning of poly-caprolactone for scaffold manufacturing: Experimental investigation on the process parameter influence. Procedia CIRP. 2016;**49**(2016):8-13. DOI: 10.1016/j.procir.2015.07.020

[5] Demir MM, Yilgor I, Yilgor E, Erman B. Electrospinning of polyurethane fibers. Polymer. 2002;**43**(11):3303-3309. DOI: 10.1016/s0032-3861(02)00136-2

[6] Zong XH, Kim K, Fang D, Ran SF, Hsiao BS, Chu B. Structure and process relationship of electrospun bioabsorbable nanofiber membranes. Polymer. 2002;**43**(16):4403-4412. DOI: 10.1016/s0032-3861(02)00275-6

[7] Jun Z, Hou H, Schaper A, Wendorff JH, Greiner A. Poly-L-lactide nanofibers by electrospinning-influence of solution viscosity and electrical conductivity on fiber diameter and fiber morphology. e-Polymers. 2003;**3**(1):1-9. DOI: 10.1515/epoly.2003.3.1.102

[8] Luzio A, Canesi EV, Bertarelli C, Caironi M. Electrospun Polymer fibers for electronic application. Materials. 2014;**7**(2):906-947. DOI: 10.3390/ma7020906

[9] Zong XH, Bien H, Chung C-Y, Yin LH, Fang DF, Hsiao BS, et al. Electrospun fine-textured scaffolds for heart tissue constructs. Biomaterials. 2005;**26**(26):5330-5338. DOI: 10.1016/j.biomaterials.2005.01.052

[10] Haghi AK. Electrospinning of Nanofibers in Textiles. 1st ed. New York: Apple Academic Press; 2011. 132 p. DOI: 10.1201/b12229

[11] Andrady AL. Science and Technology of Polymer Nanofibers. 1st ed. Hoboken, NJ, USA: John Wiley & Sons; 2007. 403 p. DOI: 10.1002/9780470229842

[12] Sukigara S, Gandhi M, Ayutsede J, Micklus M, Ko F. Regeneration of Bombyx morisilk by electrospinning – Part 1: processing parameters and geometric properties. Polymer. 2003;**44**(19):5721-5727. DOI: 10.1016/s0032-3861(03)00532-9

[13] Huang Z-M, Zhang Y-Z, Kotaki M, Ramakrishna S. A review on polymer nanofibers by electrospinning and their applications in nanocomposites. Composites Science and Technology. 2003;**63**(15):2223-2253. DOI: 10.1016/s0266-3538(03)00178-7

[14] Son WK, Youk JH, Lee TS, Park WH. The effects of solution properties and polyelectrolyte on electrospinning of ultrafine poly (ethylene oxide) fibers. Polymer. 2004;**45**(9):2959-2966. DOI: 10.1016/j.polymer.2004.03.006

[15] Ding B, Kim HY, Lee SC, Lee DR, Choi K-J. Preparation and characterization of nanoscaled poly(vinyl alcohol) fibers via electrospinning. Fibers and Polymers. 2002;**3**(2):73-79. DOI: 10.1007/bf02875403

[16] Lee JS, Choi KH, Ghim HD, Kim SS, Chun DH, Kim HY, et al.. Role of molecular weight of atactic poly(vinyl alcohol) (PVA) in the structure and properties of PVA nanofabric prepared by electrospinning. Journal of Applied Polymer Science. 2004;**93**(4):1638-1646. DOI: 10.1002/app.20602

[17] Kim JR, Choi SW, Jo SM, Lee WS, Kim BC. Characterization and properties of P(VdF-HFP)-based fibrous polymer electrolyte membrane prepared by electrospinning. Journal of the Electrochemical Society. 2005;**152**(2):A295-A299. DOI: 10.1149/1.1839531

[18] Gupta P, Elkins C, Long TE, Wilkes GL. Electrospinning of linear homopolymers of poly(methyl methacrylate): Exploring relationships between fiber formation, viscosity, molecular weight and concentration in a good solvent. Polymer. 2005;**46**(13):4799-4810. DOI: 10.1016/j.polymer.2005.04.021

[19] Stanger J, Tucker N, Staiger M. Electrospinning. 1st ed. Shropshire, United Kingdom: Smithers Rapra Technology; 2005. 206 p. DOI: 978-1-84735-091-6

[20] Jiang HL, Fang DF, Hsiao BS, Chu B, Chen WL. Optimization and characterization of dextran membranes prepared by electrospinning. Biomacromolecules. 2004;**5**(2):326-333. DOI: 10.1021/bm034345w

[21] Ki CS, Baek DH, Gang KD, Lee KH, Um IC, Park YH. Characterization of gelatin nanofiber prepared from gelatin-formic acid solution. Polymer. 2005;**46**(14):5094-5102. DOI: 10.1016/j.polymer.2005.04.040

[22] Abd Razak SI, Wahab IF, Fadil F, Dahli F, Khudzari A, Adeli H. A review of electrospun conductive polyaniline based Nanofiber composites and blends: Processing features, applications, and future directions. Materials Science and Engineering. 2015;**2015**(1):356286. DOI: 10.1155/2015/356286

[23] He J-H, Wan Y-Q, Yu J-Y. Allometric scaling and instability in electrospinning. International Journal of Nonlinear Sciences and Numerical Simulation. 2004;**5**(3):243-252. DOI: 10.1515/ijnsns.2004.5.3.243

[24] Wan Y-Q, He J-H, Wu Y, Yu J-Y. Vibrorheological effect on electrospun polyacrylonitrile (PAN) nanofibers. Materials Letters. 2006;**60**(27):3296-3300. DOI: 10.1016/j.matlet.2006.03.007

[25] Wan Y-Q, He J-H, Wu Y, Yu J-Y. Carbon nanotube-reinforced polyacrylonitrile nanofibers by vibration-electrospinning. Polymer International. 2007;**56**(11):17-26. DOI: 10.1002/pi.2358

[26] Hristian L, Ostafe MM, Manea LR, Leon AL. The study about the use of the natural fibers in composite materials. IOP Conference Series: Materials Science and Engineering. 2016;**145**(1):032004. DOI: 10.1088/1757-899x/145/3/032004

[27] McKee MG, Hunley MT, Layman JM, Long TE. Solution rheological behavior and electrospinning of cationic polyelectrolytes. Macromolecules. 2006;**39**(2):575-583. DOI: 10.1021/ma051786u

[28] Haghi AK, Zaikov GE, editors. Development of Nanotechnology in Textiles. 1st ed. New York: Nova Science Publishers, Inc.; 2012. 180 p. DOI: 978-1-62081-052-1

[29] Lee CH, Shin HJ, Cho IH, Kang Y-M, Kim IA, Park K-D, et al. Nanofiber alignment and direction of mechanical strain affect the ECM production of human ACL fibroblast. Biomaterials. 2005;**26**(11):1261-1270. DOI: 10.1016/j.biomaterials.2004.04.037

[30] Wannatong L, Sirivat A, Supaphol P. Effects of solvents on electrospun polymeric fibers: Preliminary study on polystyrene. Polymer International. 2004;**53**(11):1851-1859. DOI: 10.1002/pi.1599

[31] Yu JH, Fridrikh SV, Rutledge GC. Production of submicrometer diameter fibers by two-fluid electrospinning. Advanced Materials. 2004;**16**(17):1562-1566. DOI: 10.1002/adma.200306644

[32] Zeng J, Xu XY, Chen XS, Liang QZ, Bian XC, Yang LX, Jing XB. Biodegradable electrospun fibers for drug delivery. Journal of Controlled Release. 2003;**92**(3):227-231. DOI: 10.1016/s0168-3659(03)00372-9

[33] Choi JS, Lee SW, Jeong L, Bae SH, Min BC, Youk JH,Park WH. Effect of organosoluble salts on the nanofibrous structure of electrospun poly(3-hydroxybutyrate-co-3-hydroxyvalerate). International Journal of Biological Macromolecules. 2004;**34**(4):249-256. DOI: 10.1016/j.ijbiomac.2004.06.001

[34] Bhardwaj N, Kundu SC. Electrospinning: A fascinating fiber fabrication technique. Biotechnology Advances. 2010;**28**(3):325-347. DOI: 10.1016/j.biotechadv.2010.01.004

[35] de Vrieze S, Van Camp T, Nelvig A, Hagstrom B, Westbroek P, De Clerck K. The effect of temperature and humidity on electrospinning. Journal of Materials Science. 2008;**44**(5):1357-1362. DOI: 10.1007/s10853-008-3010-6

[36] Arumugam GK, Khan S, Heiden PA. Comparison of the effects of an ionic liquid and other salts on the properties of electrospun fibers, 2–poly (vinyl alcohol). Macromolecular Materials and Engineering. 2009;**294**(1):45-53. DOI: 10.1002/mame.200800199

[37] Pillay V, Dott C, Choonara Y, Tyagi C, Tomar L, Kumar P, et al.. A review of the effect of processing variables on the fabrication of electrospun nanofibers for drug delivery applications. Journal of Nanomaterials. 2013;**2013**(1):1-22. DOI: 10.1155/2013/789289

[38] Li Z, Wang C. One-Dimensional Nanostructures Electrospinning Technique and Unique Nanofibers. 1st ed. Berlin, Heidelberg: Springer; 2013. 141 p. DOI: 10.1007/978-3-642-36427-3

[39] Manea LR, Bertea AP, Nechita E, Popescu CV, Sandu I. Mathematical model of the electrospinning process II. Effect of the technological parameters on the electrospun fibers diameter. Revista de chimie (Bucharest). 2016;**67**(8):1607-1612. Available from: http://www.revistadechimie.ro/pdf/MANEA%20R$%208%2016.pdf

[40] Manea LR, Hristian L, Leon AL, Popa A. Recent advances of basic materials to obtain electrospun polymeric nanofibers for medical applications. IOP Conference Series: Materials Science and Engineering. 2016;**145**(1):032006. DOI: 10.1088/1757-899x/145/3/032006

[41] Maftei D, Asaftei IV, Sandu I, Manea LR, Birsa LM, Earar K. Conversion of industrial feedstock mainly with butanes and Butenes over HZSM-5 and Zn/HZSM-5 (nitrate) catalysts. Revista de chimie (Bucharest). 2015;**65**(5):673-680. Available from: http://www.revistadechimie.ro/pdf/ASAFTEI%20I.pdf%205%2015.pdf

[42] Manea LR, Cramariuc B, Caunii V, Sandu I. Equipment for obtaining polimeric nanofibers by electrospinning technology I. Constructive and functional elements of the computerized electrospinning equipment. Materiale Plastice. 2015;**52**(1):82-86. Available from: http://www.revmaterialeplastice.ro/pdf/MANEA%20L.pdf%201%2015.pdf

[43] Manea LR, Danu MC, Sandu I. Effect of the applied electric voltage and flow rate on electrospun fibers diameter. Revista de chimie (Bucharest). 2015;**66**(6):868-673. Available from: http://www.revistadechimie.ro/pdf/MANEA%20L.pdf%206%2015.pdf

[44] Manea LR, Bertea AP, Nechita E, Popescu CV, Sandu I. Mathematical model of the electrospinning process I. Effect of the distance between electrodes on the electrospun fibers diameter. Revista de chimie (Bucharest). 2016;**67**(7):1284-1289. Available from: http://www.revistadechimie.ro/pdf/MANEA%20ROZE%207%2016.pdf

[45] Manea LR, Cramariuc B, Popescu V, Cramariuc R, Sandu I, Cramariuc O. Equipment for obtaining polimeric nanofibers by electrospinning technology II. The obtaining of polimeric nanofibers. Materiale Plastice. 2015;**52**(2):180-185. Available from: http://www.revmaterialeplastice.ro/pdf/MANEA%20L.pdf%202%2015.pdf

[46] Zargham S, Bazgir S, Tavakoli A, Rashidi AS, Damerchely R. The effect of flow rate on morphology and deposition area of electrospun nylon 6 nanofiber. Journal of Engineered Fibers and Fabrics. 2012;**7**(4):42-49. Available from: http://www.jeffjournal.org

[47] Megelski S, Stephens JS, Chase DB, Rabolt JF. Micro- and nanostructured surface morphology on electrospun polymer fibers. Macromolecules. 2002;**35**(22):8456-8466. DOI: 10.1021/ma020444a

[48] Yuan X, Zhang Y, Dong C, Sheng J. Morphology of ultrafine polysulfone fibers prepared by electrospinning. Polymer International. 2004;**53**(11):1704-1710. DOI: 10.1002/pi.1538

[49] Chowdhury M, Stylios G. Effect of experimental parameters on the morphology of electrospun nylon 6 fibers. International Journal of Basic & Applied Sciences. 2010;**10**(6):70-78. DOI: 100906-2323

[50] Zafar M, Najeeb S, Khurshid Z, Vazirzadeh M, Zohai S, Najeeb B, Sefat F. Potential of electrospun nanofibers for biomedical and dental applications materials. Materials. 2016;**9**(2):73-81. DOI: 10.3390/ma9020073

[51] Yalcinkaya F, Yalcinkaya B, Jirsak O. Dependent and independent parameters of needleless electrospinning. In: Haider S, Haider A, editors. Electrospinning – Material, Techniques and Biomedical Applications. 1st ed. Rijeka, Croatia: InTech; 2016. pp. 67-93. DOI: 10.5772/65838

[52] Wang T, Kumar S. Electrospinning of polyacrylonitrile nanofibers. Journal of Applied Polymer Science. 2006;**102**(2):1023-1029. DOI: 10.1002/app.24123

[53] Matabola KP, Moutloali RM. The influence of electrospinning parameters on the morphology and diameter of poly (vinyledene fluoride) nanofibers-effect of sodium chloride. Journal of Materials Science. 2013;**48**(16):547-5482. DOI: 10.1007/s10853-013-7341-6

[54] Zhang C, Yuan X, Wu L, Han Y, Sheng J. Study on morphology of electrospun poly(vinyl alcohol) mats. European Polymer Journal. 2005;**41**(3):423-432. DOI: 10.1016/j.eurpolymj.2004.10.027

[55] Yördem OS, Papila M, Menceloğlu YZ. Effects of electrospinning parameters on polyacrylonitrile nanofiber diameter: An investigation by response surface methodology. Materials & Design. 2008;**29**(1):34-44. DOI: 10.1016/j.matdes.2006.12.013

[56] Sill TJ, von Recum HA. Biomaterials, electrospinning: applications in drug delivery and tissue engineering. Biomaterials. 2008;**29**(13):1989-2006. DOI: 10.1016/j.biomaterials.2008.01.011

[57] Supaphol P, Mit-Uppatham C, Nithitanakul M. Ultrafine electrospun polyamide-6 fibers: effects of solvent system and emitting electrode polarity on morphology and average fiber diameter. Macromolecular Materials and Engineering. 2005;**290**(9):3699-3709. DOI: 10.1002/mame.200500024

[58] Reneker DH, Yarin AL, Zussman E, Xu H. Electrospinning of nanofibers from polymer solutions and melts. In: Aref H, Vandergiessen E, editors. Advances in Applied Mechanics. Vol. 41. 1st ed. Cambridge, Massachusetts: Academic Press; 2007. pp. 43-195 (345-346). DOI: 10.1016/s0065-2156(07)41002-x

[59] Zhou L, Gong RH, Porat I. Polymeric nanofibers via flat spinneret electrospinning. Polymer Engineering & Science. 2009;**49**(12):2475-2481. DOI: 10.1002/pen.21498

[60] Ramakrishnan R, Gimbun J, Samsuri F, Narayanamurthy V, Gajendran N, Sudha Lakshmi Y, et al. Needleless electrospinning technology – An entrepreneurial perspective. Indian Journal of Science and Technology. 2016;**9**(15):2-11. DOI: 10.17485/ijst/2016/v9i15/91538

[61] Beachley V, Wen X. Effect of electrospinning parameters on the nanofiber diameter and length. Materials Science and Engineering: C. 2009;**29**(3):663-668. DOI: 10.1016/j.msec.2008.10.037

[62] Tan SH, Inai R, Kotaki M, Ramakrishn S. Systematic parameter study for ultra-fine fiber fabrication via electrospinning process. Polymer. 2005;**46**(16):6128-6134. DOI: 10.1016/j.polymer.2005.05.068

[63] Subbiah T, Bhat GS, Tock RW, Parameswaran S, Ramkumar SS. Electrospinning of nanofibers. Journal of Applied Polymer Science. 2005;**96**(2):557-569. DOI: 10.1002/app.21481

[64] Moutsatsou P, Coopman K, Smith MB, Georgiadou S. Conductive PANI fibers and determining factors for the electrospinning window. Polymer. 2015;**77**(2015):143-151. DOI: 10.1016/j.polymer.2015.08.039

[65] Rafiel S, Maghsoodloo S, Noroozi B, Mottaghitalab V, Hafgi AK. Mathematical modeling in electrospinning process. Cellulose Chemistry and Technology. 2013;**47**(5-6):323-328. Available from: http://www.cellulosechemtechnol.ro/pdf/CCT5-6(2013)/p.323-338.pdf

[66] Manea LR, Sandu I. Study concerning the processability of polyetherimide (PEI) solution for obtaining nanofibers. Revista de chimie (Bucharest). 2015;**66**(12):1968-1973. Available from: http://www.revistadechimie.ro/pdf/MANEA%2012%2015.pdf

[67] Manea LR, Scârlet R, Amariei N, Nechita E, Sandu IG. Study on behaviour of polymer solutions in electrospinning technology. Revista de chimie (Bucharest). 2015;**66**(4):542-546. Available from: http://www.revistadechimie.ro/pdf/MANEA%20L.pdf%204%2015.pdf

Chapter 7

Development of New Nanostructured Electrodes for Electrochemical Conversion: Energy and Fuels from the Environment

Giulia Massaglia and Marzia Quaglio

Additional information is available at the end of the chapter

http://dx.doi.org/10.5772/intechopen.75352

Abstract

During the last decade, carbon-based nanofibers emerged as an important class of materials for the fabrication of electrodes for electrochemical energy conversion. Indeed carbon-based nanofibers combine high electrochemical stability and high porosity to high mechanical flexibility and low weight, resulting in a unique and versatile material for the design and fabrication of energy-related devices. This chapter aims to show and analyze new nanostructured materials, obtained by electrospinning technique, in order to design 3D arrangement of the electrodes and to improve the energy efficiency of energy production devices. Indeed, the design of new 3D nanostructured electrodes enhances the energy efficiency of these devices, optimizing the energy production, obtained by new renewable energy technologies. The chapter is focused on those devices able to generate power output through the electrochemical conversion of different fuels, like wastes, and environmental compounds, such as CO_2.

Keywords: carbon nanofibers, electrospinning, carbon based electrodes, energy conversion devices, energy storage devices

1. Introduction

Electrospinning is a unique process for the fabrication of 1D nanomaterials. During the last decades, it gained progressive interest, demonstrating huge potential in the different scientific areas. Especially in the field of energy and electrochemical energy conversion (EEC), electrospinning has a leading role among the techniques for the fabrication of nanomaterials.

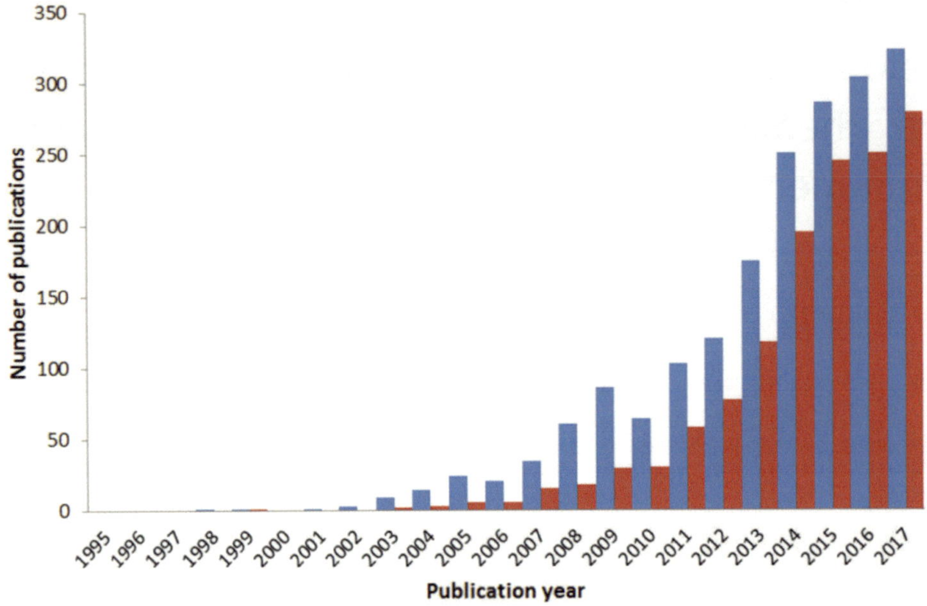

Figure 1. Number of publications discussing carbon-based nanofibers by electrospinning in blue, and in red the number of works specifically related to energy is reported (source SCOPUS).

Indeed, it offers the possibility to fabricate 1D nanostructures with good control of the nanofiber morphology and of their arrangement in the final mats.

In EEC devices, carbon-based nanofibers have attracted particular interest for the design of high-performing electrodes. They combine high electrochemical stability to unique mechanical properties, exhibiting high surface area to volume ratio.

The high interest on carbon-based nanofibers by electrospinning is evidently analyzed in **Figure 1**, in which the number of works on this topic, published during the last 20 years, is reported by the blue columns. What is even more interesting is that a significant number of those works discussed the application in energy of carbon nanofibers, as described by the red columns.

This chapter analyzes the fabrication of carbon-based nanofibers by electrospinning, discussing their formation from different carbon precursors proposed in the literature, presenting the processing of nanofibrous electrodes for EEC and reporting key examples of their application in different EEC systems.

2. Electrochemical energy conversion

Electrochemical energy conversion (EEC) refers to the conversion of chemical energy into electrical energy by the proper control of reduction-oxidation (redox) reactions. By processes of this kind, it is possible to convert chemical energy, trapped in chemical bonds of different molecules acting like fuels, into electrical energy as in fuel cells (FC) [1], or to harvest solar energy of photons thanks to the presence of proper molecules decorating the surface of a

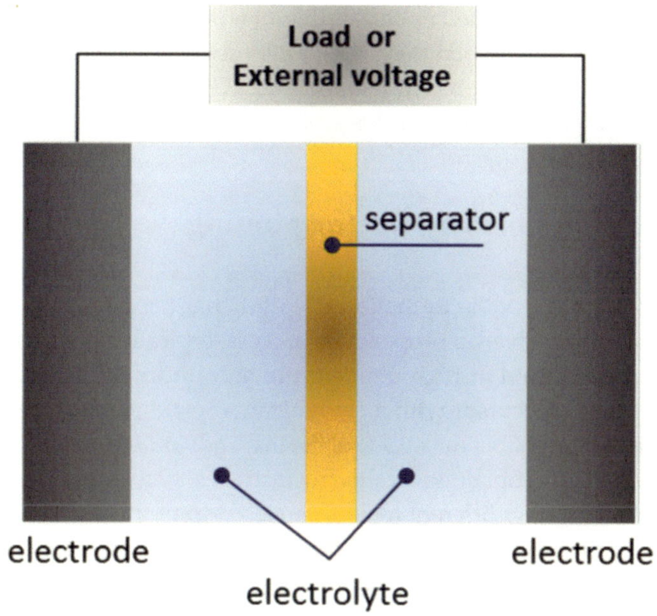

Figure 2. Schematic representation of an electrochemical reactor.

semiconducting oxide as in dye-sensitized solar cells (DSSC) [2]. EEC also permits to store energy for its further use as in supercapacitors, by the creation of a double-layered charges, [3, 4] and in Li-ion batteries, by the so-called intercalation process [5]. Moreover, EEC can be associated to redox reactions, induced to obtain new molecules, able to efficiently store the starting energy into their chemical bonds. Interesting examples are electrocatalytic water splitting for H_2 production, [6] or CO_2 photoelectrochemical reduction [7].

From a general point of view, all these electrochemical reactors have a common structure, as reported in **Figure 2**. Indeed, they all are made of two electrodes, that is, anode and cathode, a liquid (or semisolid) electrolyte, eventually containing a separator. Actually, the optimization of the different EEC systems requires specific strategies to properly control the crucial reactions and processes occurring inside them. General agreement exists in this field on the key role played by the material constituting the electrodes. For each EEC system, the electrode material needs to satisfy some requirements. It must

- be chemically robust, so as to sustain the electrochemical reaction for long time without degrading or limiting the reaction efficiency;
- be high electrically conductive;
- expose high surface area to volume ratio, so expose high area to favor the reactions;
- have high durability over time;
- have high mechanical strength, possibly combined to good mechanical flexibility.

During the last decade, carbon-based materials demonstrated to be the best candidate to fulfill all the requirements, especially when they are nanostructured. In this scenario, carbon-based nanofibers have started to be explored in the field of EEC, showing a tremendous potential to contribute to the further development of this area.

3. Carbon-based nanofibers by electrospinning

Electrospinning process ensures the formation of carbon nanofibers, starting from a polymeric solution and applying successively a proper post-process, typically a thermal treatment, known as pyrolysis process, conducted at high temperature and under an inert atmosphere [8]. The selection of the polymeric precursor with a carbonization yield plays a crucial role in order to obtain final carbon-based nanofiber mats. One of the main advantages, offered by electrospinning technique, is the possibility to obtain several nanostructures, such as hollow nanofibers, porous and dense nanofibers, by using different tools. The electrospinning setup is basically characterized by three parts: (1) the spinneret that hosts the needle representing the first electrode; (2) the counter electrode, also named collector, which is the second electrode and it ensures the collection of dried nanofiber mats; and (3) high voltage supply [9–12]. As sketched in **Figure 3**, the high

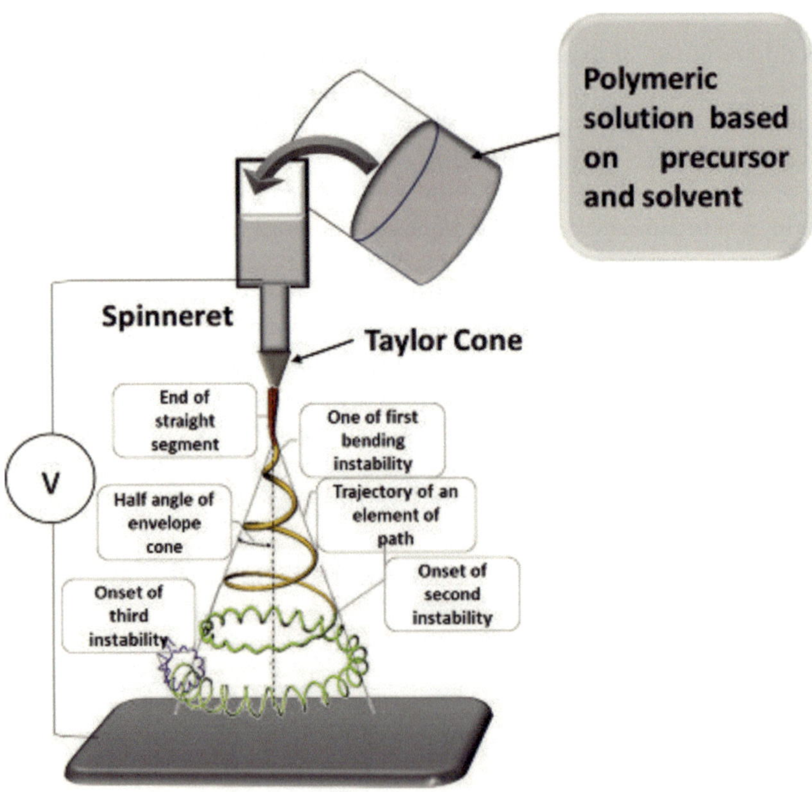

Figure 3. Scheme of the electrospinning process together with the representation of bending instabilities achieved during the process. Reprinted with the permission from (Polymer, 2008, 49, 2387–2425) Copyright (2008) Elsevier.

voltage (in the range between 0 and 30 kV) is applied to induce charges distribution inside the polymeric drop that is shaped at the tip of the needle. The interaction between all charges generates a repulsive force, which typically increases as the voltage value grows. Indeed, as voltage progressively increases, the spherical shape of the drop is stretched, assuming a conical shape, named Taylor cone. When the applied voltage overcomes the threshold value, the repulsive force overcomes the surface tension of solution, inducing the ejection of charged polymeric jet.

Since during the process, both bending instabilities and solvent evaporation induce the stretching of the charged polymeric jet, leading to provide the deposition of nanofibers with diameters in the order from few nanometers to some micrometers, the definition of working distance is quite important [13–15]. The working distance is defined as the distance between the tip of needle and the counter electrode and it is directly correlated with the possibility to collect on the counter electrode a dried nanofiber mat. All the process parameters play a crucial role to tune the morphological properties of the nanofibers. Furthermore, the subdivision of these parameters in three main classes is completely known:

i. Polymeric solution parameters, such as viscosity, electrical conductivity, surface tension, concentration and polymer molecular weight;

ii. Process parameters, as applied voltage, working distance, flow rate and strength electric field, defined as the ratio between the voltage and the working distance;

iii. Environmental parameters, such as room temperature and humidity.

Moreover, the spinnability of the polymeric solution is strictly correlated with the solution parameters. In order to obtain the formation of a continuous charged polymeric jet during the process, the solution viscosity must be in the range (Eq. 1) [16–18]:

$$0.02 \le \eta \le 200\ Pa * s \tag{1}$$

When the viscosity value, (η (Pa*s)), is lower than 0.1 Pa*s, charged droplets are formed during the process, giving rise to electrospray process. On the contrary, when the solution viscosity is higher than 2 Pa*s, the formation of a continuous charged polymeric jet occurred, leading thus to collect dried nanofiber mats. Different works in the literature demonstrate the direct correlation between the solution viscosity and the uniform morphological properties, characterizing the nanofiber mat [16]. It is possible to define a direct correlation between polymer concentration and solution viscosity: higher is the polymer concentration, higher is the solution viscosity. When the polymer concentration is low, thinner nanofiber mats are collected on the counter electrode.

Regarding the second class of parameters, the flow rate and working distance affect the morphological properties of final nanofibers. High flow rate values, indeed, induce the formation of nanofibers, characterized by a no uniform diameter distribution and by a large number of defects. Whereas low flow rate values cause the formation of the Taylor cone inside the needle, leading to the formation of no continuous charged polymeric jet and the collection of beads nanostructured into the mats. Since the evaporation of the solvent can depend on the working distance, its value influences the collection of the final dried nanofiber mats.

3.1. The reference process: Poly (acrylonitrile) as the carbon source

As deeply investigated in the literature [8, 19], the main polymeric precursor used in order to obtain carbon-based nanofibers is Poly (acrylonitrile) (PAN). The polymer chain of PAN is represented in **Figure 4**.

PAN is selected as precursor thanks to its properties, such as high carbonization yield, high melting polymer and high content of nitrogen, leading to the self-induced nitrogen doping into the final nanofiber mats, modulating properly the heating treatment [19]. Different works in the literature investigate the role of nitrogen in order to optimize the carbon-based nanofibers, designing new electrode to improve the overall performance of electrochemical devices. In particular, the so-called activated carbon-based nanofibers (ACNFs) are obtained by applying successive chemical treatments (e.g. ammonia treatments, oxidation treatment in nitric acid and so on) on carbon nanofibers after the thermal treatment conducted at temperatures higher than 100°C [20]. Whereas, PAN nanofibers can show a self-induced nitrogen doping, when the pyrolysis treatment is conducted at low temperature values, in the range from 600–900°C. In order to obtain a final polymeric solution suitable for electrospinning, the most common polymeric mixture is based on PAN dissolved in an organic solvent, as Dimethylformamide (N-N DMF). Different works in the literature investigate the correlation between all involved electrospinning parameters and the morphological properties of PAN nanofibers [8, 21]. Therefore, Yordem et al. [8] demonstrate that the working distance results to be the main parameter, which can be influenced by the diameter distribution in the range of several nanometers. PAN nanofibers can be obtained by starting from a polymeric solution, containing a low PAN concentration (in the range 8–10 wt%) and applying a voltage value among 10–20 kV.

The pyrolysis treatment is the heat treatment, carried out in order to transform the PAN nanofiber mats in carbon nanofibers (CNFs). Liu et al. [22] showed the pyrolysis treatment, divided into three main steps: *(i)* the oxidative stabilization; *(ii)* the carbonization; and *(iii)* the graphitization. All these steps should be implemented in order to maintain the nanostructures during the conversion of PAN fibers into CNFs. Among all these steps, the most important and complex stage results to be the oxidative stabilization. This step plays a crucial role in the definition of carbon nanofiber structures and involves several chemical reactions, such as cyclization, dehydrogenation, aromatization, oxidation and crosslinking [23, 24]. The oxidative stabilization occurred in air at temperature lower or equal to 280°C and during this step, the $C \equiv N$ bonds, characterizing the polymeric chains, is turned into $C = N$, leading thus to the

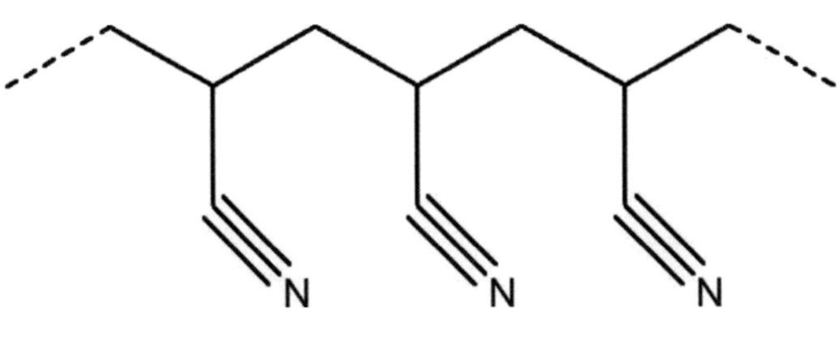

Figure 4. Molecular chain of polymer polyacrylonitrile.

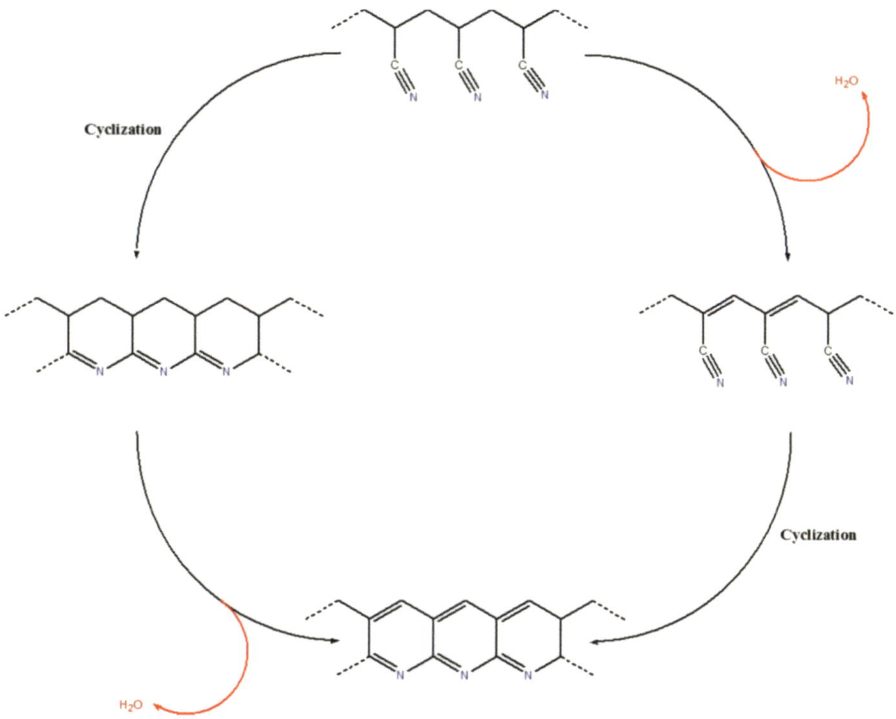

Figure 5. Scheme of the two main reactions occurred during the oxidative stabilization: Cyclization and dehydrogenation, which induces the formation of a water molecule.

crosslinking between PAN molecules and the thermal stabilization of the nanostructures [25]. The main reactions, occurred during this step, are dehydrogenation and cyclization reactions, as sketched in **Figure 5**.

In particular, the dehydrogenation involves the formation of double bonds between nitrogen and carbon atoms and successively the removal of water molecules. Whereas the cyclization is the reaction able to create the ring configuration in the main chain. Indeed, the nitrile groups react with each adjacent group, originating then stable polymeric structure. Once obtained a nanostructured material thermally stable, the following steps of carbonization and graphitization are carried out. Typically, both of the two steps are conducted under inert atmosphere (using argon or nitrogen flow) at high temperature values. The carbonization step is occurred at temperature higher than 900°C and during this phase, the ring structure starts to arrange itself to get the formation of small size graphene sheets. In particular, a molecule of HCN broke out and a reduction of nitrogen content in the main chain takes place. The last graphitization step is conducted for temperature higher than 1000°C (1000°C < T < 3000°C) in order to convert the majority of PAN precursor into a carbon structure, ensuring thus the formation of larger graphitic ordered sheets. Although the formation of uniform CNFs mats, starting from PAN as precursor, results to be difficult, the final CNFs show unusual properties, thanks to their high surface area ratio to volume and high porosity webs, making them suitable to design new nanostructured electrodes.

3.2. Other carbon precursors

The class of polymers, reported in **Table 1**, which can be used as precursor in order to obtain carbon-based nanofibers, turn out to be restricted. Theoretically, the carbon backbone ensures the possibility to convert that polymer precursor in a carbon-based material.

One of the most interesting possibilities is the selection of a natural polysaccharide, such as cellulose, chosen as carbon precursor. Deng et al. [26] fabricated CNFs by using nano-sized cellulosic precursors. The nanofibers, obtained by electrospinning of a polymeric solution based on cellulose acetate, were then left in a solution of 0.5 M NaOH dissolved in ethanol to obtain regenerate cellulose-based nanofibers. Subsequently, a pyrolysis treatment was carried out through two steps, converting thus cellulose-based nanofibers into CNFs. The first step is conducted in air at 240°C in order to stabilize the nanostructure, while the carbonization (second step) is obtained by varying the temperature in the range from 800–2200°C under argon flow. In this work, it was possible to observe that the diameters decrease with the increasing of the heating temperature. In particular, the diameters varied from 430 to 200 nm, when the heating temperature is close to 2200°C. These obtained CNFs show an improvement of mechanical resistances, due to their high surface area and small diameter distribution. Moreover, the yield of carbonization of cellulose is ensured by the possibility to obtain a graphitic-like nanostructure starting from 1500°C.

Different synthetic polymers can be used as precursors for carbonization: polyimide (PI), poly(vinyl alcohol) (PVA) and poly(vinyliden fluoride) (PVDF). Many works in the literature investigate the formation of PI nanofibers [27–30], by providing three different steps: (i) preparation of polymeric solution based on polyamic acid (PAA); (ii) electrospinning of this polymeric solution and (iii) imidization of the PAA nanofiber mats. The imidization process is carried out in N_2 atmosphere; during this process two heating phases are implemented: the first is conducted at 150°C for 40 min (heating rate equal to 5°C min^{-1}) and the second one at 280°C for 40 min (heating rate is 2°C min^{-1}) [27]. The carbonization process was conducted at a temperature of 1000°C with a heating rate of 10°C min^{-1} in inert atmosphere [30, 31]. Different approaches were investigated in order to induce the graphitization of the samples [31] and/or to create a N_2 doping to functionalize the nanofiber mats [30]. Indeed, Yang et al. [31] sandwiched the carbonized PI nanofibers into graphite plates and treated them at 2200°C in He atmosphere. Whereas Kim et al. [30], in order to activate the carbon-based nanofibers after the carbonization process and thus optimize the materials as supercapacitors, implemented a heating treatment at temperature in the range from 650–800°C with a 40%vol steam in the nitrogen. PAA concentration or viscosity solution are the two main parameters that can influence the morphological properties of PI nanofibers [32], as represented in **Figure 6**. The diameters of PI nanofibers vary in the range between some tens and several hundreds nanometers [8]. In order to tune the morphological properties of final carbon nanofibers and their porosity distribution, it is important to underline that the diameter values decrease during the imidization process and the carbonization treatment [31].

The thermoplastic polymer PVDF is used as carbon precursor thanks to its intrinsic properties, which ensure the formation of a continuous charged polymeric jet during the electrospinning, avoiding as much as possible the instauration of nano-droplets during the process [33].

Electrospun carbon-based nanofibers			
Polymeric precursors	**Pyrolysis process**	**Parameters involved in nanofibers morphology**	**References**
PAN dissolved in N-N DMF, used as solvent	Oxidative stabilization conducted in air at T = 280°C Carbonization step conducted under inert atmosphere at T ≥ 900°C Graphitization step conducted under inert atmosphere at 100°C < T < 3000°C	Working distance Polymer concentration Voltage applied in the range from 10 to 20 kV	[8, 20–25]
Nano-sized Cellulosic Precursor	In air at T = 240°C Carbonization step conducted under inert atmosphere at two temperatures: T1 = 800°C and T2 = 2200°C	Heating temperature: average diameters vary in the range from 430 to 200 nm;	[26]
PI nanofibers obtained by starting from a solution based on PAA	Imidization process: in N_2 T_1 = 150°C(50 min);T_2 = 280°C (40 min) Carbonization step::inert atmosphere T = 1000°C	PAA concentration Viscosity solution The diameters of PI nanofibers decrease during imidization process	[8, 27–31]
PVDF nanofibers	Oxidative stabilization process in air at T = 400°C Carbonization process in N2 flow is conducted at T = 1000°C	Polymer concentration: higher is the polymeric concentration, higher is the diameter values	[33]
Porous nanofibers obtained starting from a polymeric solution based on PVDF and PEO dissolved in DMF	Oxidative stabilization process in air at T = 400°C Carbonization process in N2 flow is conducted at T = 1000°C	Higher is the molecular weight of PEO and its concentration, higher is the pore distribution inside the mat and on nanofibers surface.	[34]
PVA nanofibers	Carbonization process is implemented at low temperature T = 500°C under flow of argon and hydrogen	pH of polymeric solution: diameters increase as the basic pH increases; pH is more acid the formation of charged droplets occurred	[35, 36]
Core-shell and hollow carbon-based Nanofibers			
Polymeric precursors	**Pyrolysis process**	**Hollow Carbon-based nanofibers**	**References**
PAN dissolved in N-N DMF, used as shell PMMA, mixed in a solvent of DMF/acetone, used as core	Oxidative stabilization conducted in air at T = 250°C Carbonization step conducted under inert atmosphere at T = 1100°C	Hollow nanofibers show an external diameter equal to 7 μm and the internal diameter is close to 2 μm	[37–39]

Table 1. Summary of process parameters involved during the electrospinning process and the pyrolysis process, in order to establish a correlation with the morphological properties of carbon-based nanofibers.

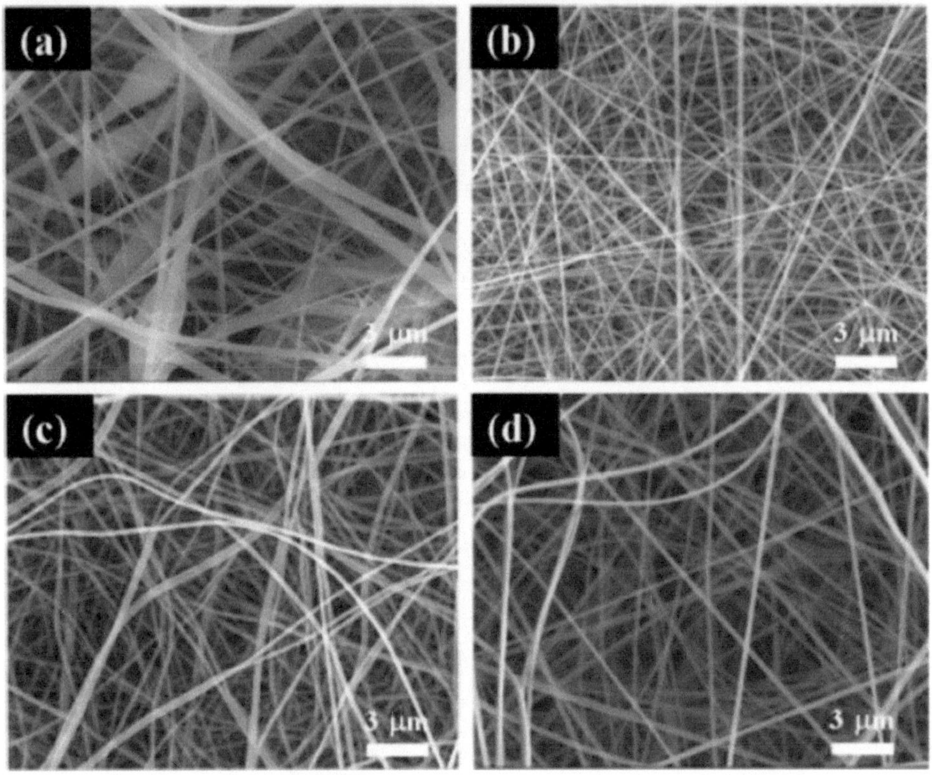

Figure 6. Diameters shrinkage provided/reached in (a) PAA nanofibers; (b) PI nanofibers after the imidization process and (c) and (d) after the carbonization process, conducted at 1000°C for 1 h. Reprinted with the permission from (journal of physical chemistry B letters, 2007, 111, 11,350–11,353) copyright (2007) American Chemical Society.

Kim et al. investigate the correlation between the morphological properties of mats with the polymer concentration. Indeed, different electrospun polymeric solutions were prepared by dissolving 12, 14, 16 and 18 wt% of PVDF in a mixture, based on acetone and dimethylacetamide (DMAc, volume ratio of 7/3 weight). PVDF nanofibers are characterized by an average diameter close to several hundreds nanometers and a microporous structure, defined by the interconnections between all nanofibers. As confirmed theoretically, it is possible to observe that the diameter increases with the increasing of polymer concentration: higher is the polymeric concentration, higher are the diameter values.

In order to obtain highly porous carbon-based nanofiber mats, derived from PVDF nanofibers, Yang et al. [34] synthetized PVDF nanofibers, starting from a polymeric solution, containing PEO (0.06 g) and PVDF (1 g) dissolved in 9 g of mixture of DMF and deionized water (5:3 volume ratio). The addition of PEO together with water as non-solvent for PVDF guarantees the formation of microporous structure on PVDF nanofibers surface. The heating treatment, selected in order to convert the PVDF precursor into carbon, is based on two different steps, named dehydrofluorination and carbonization. Dehydrofluorination enhances the thermal stability of the material before the carbonization step. Different from the thermal stability conducted for PAN nanofibers, which is achieved at low temperature close to 300°C, for PVDF the stabilization

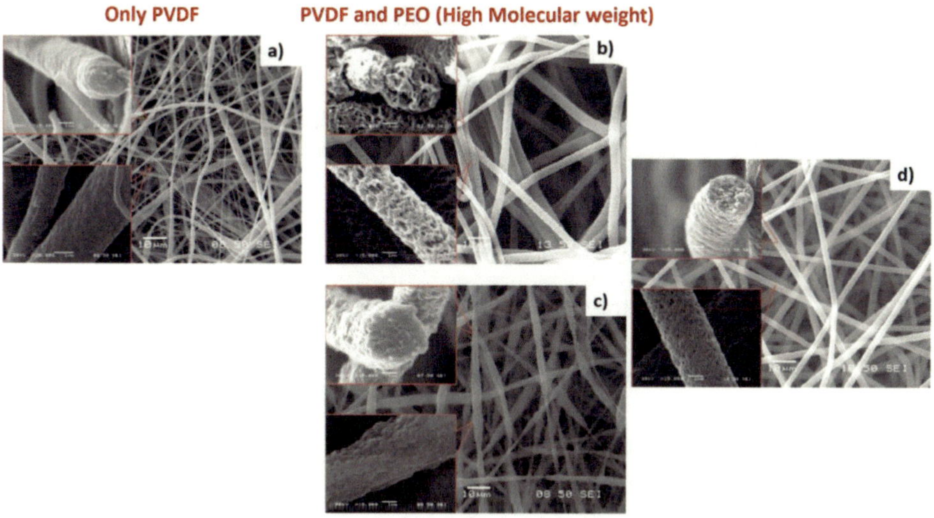

Figure 7. Modulation of porous structure in the PVDF nanofibers, obtained by adding PEO and increasing the relative humidity. (a) Only PVDF-based nanofibers; (b) PVDF and PEO nanofibers collected when the relative humidity is 45%; (c) relative humidity is 35%; (d) PVDF and PEO nanofibers obtained with a lower PEO concentration. Reprinted with the permission from (carbon, 2011, 49, 3395–3403) copyright (2011) Elsevier.

occurred through a heating treatment at 400°C in air. Since this temperature is much above the temperature of melting point (190°C), dehydrofluorination is commonly provided as chemical treatment at low temperature, introducing a large number of C–C bonds into the main chains. The PVDF nanofibers are soaked in the solution containing DMF and methanol (9:1 volume ratio) as solvent, and the chemical compound (DBU: 1,8-diazabicyclo[5.4.0]undec-7-ene) that is added for each unit of vinylidene fluoride unit. The carbonization process was carried out at temperature higher than 1000°C in N_2 flow for 1 h. The PEO concentration, the amount of water in the polymeric solution and the relative humidity reached when the nanofibers collected play a crucial role to control the porous morphology of the sample, as represented in **Figure 7**. When PEO is absent in the initial polymeric solution (**Figure 7a**), PVDF nanofibers show a certain surface roughness and few pores inside the mat. Whereas, high relative humidity combined with PEO provide the formation of pores on the nanofiber surface (**Figure 7b**). Both of the presence of water and PEO induce the pores distributions all through inside the fibers (**Figure 7b**, **c** and **d**); one pore is connected with each other. It is possible to define a correlation between the pores distributions (on the nanofibers surface and on the bulk) with PEO molecular weight and PEO concentration (**Figure 7d**). Indeed higher is the molecular weight of PEO, higher is no-solubility of PVDF, leading thus to increase the porous structure inside the mat and on nanofibers surface. The same trend can be observed when PEO concentration increases: the pores number on the surface is larger when PEO concentration increases.

Poly(vinyl alcohol) (PVA) is used to obtain carbon nanofibers, starting from named green polymeric solution, although its low carbonization yield and low decomposition temperature. PVA nanofibers, indeed, are obtained starting from a water-based solution and the carbonization process is implemented at low temperature, close to 500°C, for 3 h under flow of argon and hydrogen [35]. The main parameter, which influences the diameter distribution of

PVA nanofibers, is the pH of polymeric solution. At neutral pH value, the diameters are in the order of hundreds nanometers [36]; while the average diameters increase as the basic pH increases. On the contrary, when the pH becomes more acid a no continuous polymeric jet is guaranteed during the electrospinning process, inducing the formation of charged droplets

4. Carbon nanofibers morphology and composition

Considering PAN as the most suitable carbon precursor, it is possible to obtain different carbon nanofibers morphology, for example, core-shell nanofibers, hollow nanofibers, porous nanofibers, as summarized in **Table 1**.

4.1. Core-shell and hollow nanofibers

These kind of nanofibers is obtained by modifying the electrospinning setup into a coaxial electrospinning. Coaxial electrospinning is carried out by a concentric disposition of two syringe holders, where two spinning solutions can be loaded, as represented in **Figure 8**.

Core-shell nanofibers are made of a shell, typically natural or synthetic polymers, and by a core that can be a solvent or a polymer, known as sacrificial polymer. In this latter configuration, through a post-process, such as heating treatment and/or chemical treatment,

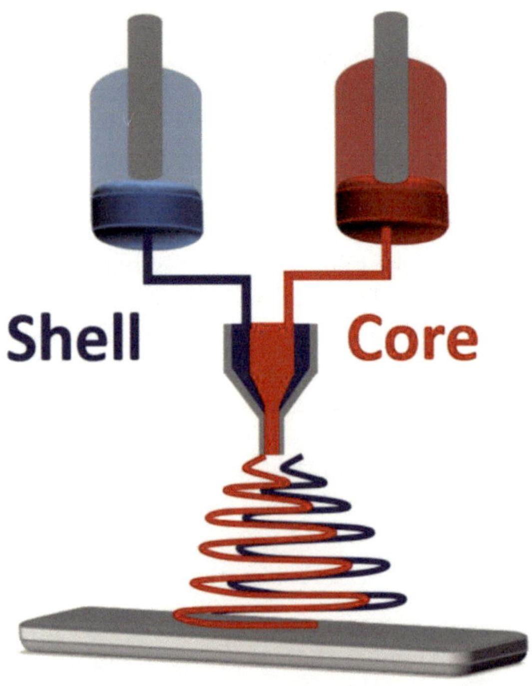

Figure 8. Sketch of coaxial electrospinning used in order to obtain core-shell nanofibers or hollow nanofibers.

the core was removed, leading thus to the formation of hollow nanofibers [37]. Therefore hollow nanofibers show an empty core and a shell, also defined wall, based on polymer, ceramic or carbon-based materials. Zussman et al. used coaxial electrospinning in order to obtain hollow carbon-based nanofibers. In this work, the shell solution contained 12 wt% of PAN dissolved in DMF, selected as carbon precursor; while the core is obtained starting from a solution of Poly(methyl)methacrylate (PMMA) mixed in a solvent of DMF/acetone. The acetone is no solvent for PAN, leading thus to create, during electrospinning process, a solid interface between shell and core and consequently a dried core-shell nanostructure. Core-shell nanofibers mat was thermally treated in order to ensure the carbonization of shell (PAN nanostructures) and simultaneously the completely degradation and decomposition of PMMA core. The heating treatment is achieved through two steps: the first step is in air at 250° to thermal stabilize the sample and the second step is in inert atmosphere (N_2 atmosphere) at 1100°C. These hollow nanofibers show an external diameter equal to 7 μm and an internal diameter close to 2 μm.

Another way, provided to obtain hollow nanofibers, is based on the preparation of electrospun solution similar to an emulsion mixture [38, 39]. Kim et al. [38] prepared the emulsion-like solution mixed two immiscible polymers: PAN as carbon source, forming the continuous phase of solution, and PMMA, which constitutes the dispersed phase. The collected nanofiber mats results to be a core-shell nanostructure: the shell is made of continuous and long fibers of PAN, whereas the core is discontinuous part of PMMA. The carbonization process ensures the formation of hollow carbon-based nanofibers through the completely degradation of PMMA polymers.

5. Application of carbon nanofibers for energy conversion

The great importance to identify new models to make human development sustainable for the environment has pushed and potentiated the scientific research in the area of renewable energy sources. Nanotechnologies are gaining a prominent role in driving this revolution toward sustainability. Indeed nanomaterials offer several advantages with respect to their macroscopic counterparts. First of all, since they offer high specific surface area combined to outstanding mechanical and electrical properties, they grant the design of high-performing devices [32, 33]. Among the different nanostructures that have been proposed in the area of EEC, nanofibers by electrospinning belong to one of the most versatile class of nanomaterials, able to be easily optimized with different morphologies and with a set of final properties that can be tuned as required by the final application [34, 35]. The development of carbon nanofibers, as electrodes in EEC devices, ensures great improvements of their overall performances [32–35]. In particular, all EEC devices described in this book chapter represent the most promising technologies in terms of renewable energy. In the last decades, the new renewable sources were developed in order to supply the 50% of energy demand, minimizing greenhouse gas emissions (GHG), limiting environmental pollution and producing power through electrochemical conversion of new fuels, such as wastes, CO_2 and other compounds.

5.1. Energy production

5.1.1. Fuel cells

Fuel cells represent an important class of electrochemical devices for energy conversion. They are capable of directly convert the chemical energy present in a wide range of molecules into electrical energy. The starting molecules, as hydrogen and methanol, act as fuel and, thanks to presence of the proper catalysts, oxidation reactions occur at the anode of the systems, resulting in the production of electricity [1]. Actually different fuel cell technologies have been developed, that make possible to use as starting fuels not only small size molecules as hydrogen, methanol and ethanol, but also more complex solid organic matter present in different king of wastewaters and soils as in happen in microbial fuel cells (MFCs) [39]. The use of carbon-based nanofibers is quite frequent in this area, since they combine good electrochemical stability with high electrical conductivity, while offering several strategies for their decoration and coupling with the required catalysts. In microbial fuel cell, particular class of microorganisms, named exoelectrogenic, catalyzes the oxidation reaction, which permits the conversion of chemical energy of organic matter of wastewater into electricity. These kinds of bacteria are able to directly release electrons that can be accepted by the anode of the device, during their metabolic activity. The electrode plays a role similar to that of different minerals that exoelectrogenic bacteria can find in their natural environments, for example, freshwater and seawater sediments [40]. Carbon-based nanomaterials are intriguing materials for the fabrication of anodes for MFCs, since they offer good electrochemical behavior and optimal morphological features that can favor bacteria growth on them [41].

An interesting possibility to improve the performances of FCs is the use of oxygen at the cathode of the system as the final electron acceptor. The reaction that occurs in these reactors is the so-called oxygen reduction reaction (ORR). The complete ORR is proposed in Eq. (2), it permits to use 4 electrons for each molecule of reacting oxygen.

$$O_2 + 4H^+ + 4e^- \rightarrow 2H_2O \tag{2}$$

Actually to promote the ORR reaction according to the pathway described by Eq. (2), avoiding the formation of dangerous and unwanted side products as H_2O_2, the presence of a proper catalyst is mandatory. The most important catalyst for this reaction is Pt, which is a rare and expensive metal [42]. Many efforts are spent worldwide to identify and optimize substitutes for Pt-based catalysts, able to offer the same catalytic efficiency but with a significant cost reduction.

Carbon-based nanofibers by electrospinning offer an interesting opportunity to design high-performing cathodes. Indeed selecting the proper precursors for the fabrication of nanofiber it is possible to add spontaneous doping sites made of heteroatoms that have demonstrated to be quite active toward the ORR. In paragraph 3.1 PAN has been introduced as the reference precursor for the synthesis of carbon-based nanofibers, due to its high carbon yield during the pyrolysis process. The structure of the polymer chain of PAN was proposed in **Figure 5**, showing the presence of -C≡N groups along the chain. Properly controlling the thermal treatment, it is possible to fabricate nanofibers with a good degree of graphitization in which several N-based defects can be present into the graphitic structure of carbon, as proposed in

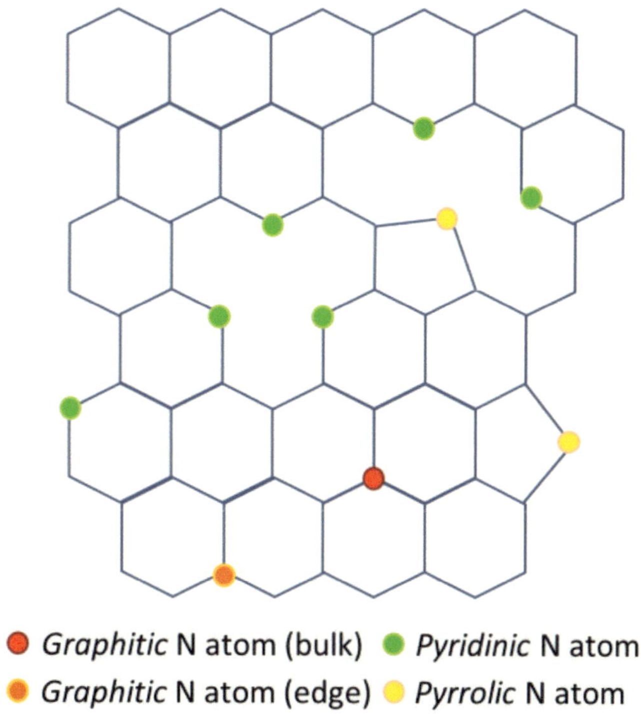

Figure 9. Sketch of a graphitic plane with the different nitrogen defects that can be formed during the pyrolysis process starting from PAN precursor.

Figure 9. It has been reported by several works in the literature that those defects actually behave as N-doping sites for the carbon nanofibers, and their presence can be controlled during the fabrication process [43]. The N-based defects in the graphitic structure of the nanofibers play an important role as active catalytic sites toward the ORR. Moreover several strategies are possible to decorate the carbon-based nanofibers with metal-based catalyst, with the possibility to optimize a co-catalysis process.

5.1.2. DSSC

Photovoltaic devices whose working principle involves electrochemical reactions have been proposed, they are the dye-sensitized solar cells (DSSCs). As in traditional solar cells, photons are directly converted in electrons, but in DSSCs organic molecules, that is, the dye, are responsible for the adsorption/conversion processes [36, 37]. A wide band gap semiconductor, which is usually TiO_2, captures the produced electrons and the regeneration of the dye is grant by the presence of a redox mediator into the electrolyte. The most frequently used redox couple is iodide/triiodide (I^-/I^{3-}), that is then reduced at the counter electrode (CE). The design of the CE is particularly challenging, since it must show high chemical resistance to the aggressive redox couple used in DSSC and preserve high efficiency over time in catalyzing the triiodide reduction. Platinum is the reference catalyst employed at the CE in DSSC, but high is the interest to substitute it with high-performing materials, leading thus to keep low the cost of the final devices and overcome the issues related to Pt that rapidly degrades because of the exposure to iodide/triiodide. The use of carbon-based nanofibers in this area has been especially important for the optimization of new counter electrodes [44, 45]. High efficiency of DSSC are strictly related to the proper design of photo-anodes, especially related to the different interfaces among all materials, that is, dye/semiconductor/electrode. Indeed the optimization of charge injection after their photo-generation is a key step to avoid charge recombination. Electrospinning offers interesting possibilities to optimize these interfaces by the preparation of composite nanofibers. An interesting example is the work by Hieu et al. [46], in which the authors optimize a photo-anode made of core/shell nanofibers with a core made of highly graphitized carbon and TiO_2 as the outer shell. The nanofibers were obtained with a starting solution made of PAN as the carbon precursor and titanium isopropoxide added to polyvinylpyrrolidone (PVP) as the TiO_2 source. The resulting DSSC performed extremely well, reaching an efficiency of 7.5%.

5.2. Energy storage

5.2.1. Lithium ion batteries

Lithium ion batteries (LIBs) are a key technology for energy storage. A LIB is made of negative and positive electrodes that can both intercalate Li^+ ions reversibly. The electrodes are separated by conducting non-aqueous electrolyte containing lithium ions. Discharge correspond to use of the battery, during this phase Li^+ ions grant the current flow from the negative to the positive electrode. The reverse operation, called charge, requires the use of an external voltage. Under the external potential, lithium ions are forced to move from the positive electrode to the negative one. At the negative electrode, lithium ions are trapped into the porous material forming the electrode during a process named intercalation [47]. Good examples, which show the use of nanofibers by electrospinning in LIBs, are especially related to the preparation of anodes, where the intercalation process causes huge mechanical stress to the materials, usually limiting their durability. The possibility to decorate electrospun carbon-based nanofibers with metal oxides is an intriguing strategy to reduce the size of the metal oxide down to the nanoscale, significantly improving the mechanical robustness of the material. The synthesis proposed by L. Ji et al. [48], is relatively easy. The authors obtained the starting polymer solution based on the addition of the oxide precursor into the solution already containing the carbon precursor. They synthetized carbon nanofibers decorated with α-Fe_2O_3 nanoparticles, demonstrating homogenous dispersion of the nanoparticles along the carbon-based nanofibers. The thermal process is carefully optimized in order to proper control the carbonization of the nanofibers, and at the same time favor the nucleation of the semiconducting oxide nanoparticles. The resulting nanofibrous composite anode showed good performances, especially in terms of its reversibility. Another interesting oxide used in LIBs anodes is SnO_x. Unfortunately, despite the huge potential of this material, the problem of its stability over time, due to the effect of the intercalation process, make impossible its real use. New approaches and processes are investigated to reduce the effect of the volume variation caused by Li ions intercalation. The fabrication of nanostructures usually helps to alleviate the problem of volume variation. Zhu et al. [49] demonstrated the possibility to co-synthetized SnOx nanoparticles directly on 1D carbon-based nanofibers, through several phases: (i) an electrospinning step, (ii) a calcination process. The resulting anodes showed extraordinary good cycling durability. The development on smart systems and new portable electronic tools has required the development of devices for energy storage able to couple light-weight to small dimensions and frequently to high mechanical flexibility. Carbon-based nanofibers represent a unique opportunity for electrochemical energy storage for the design and optimization of flexible systems. It is indeed quite interesting to consider that carbon-based nanofiber mats obtained by electrospinning usually preserve high flexibility and bending ability after the pyrolysis treatment. So that processing the nanofiber mat in such a way to have it as a freestanding membrane, it is possible to fabricate incredibly flexible electrodes that can be integrated in LIBs. As an example the work of Samul et al. [50] can be considered. They demonstrated that carbon-based nanofibers were able to preserve their high flexibility even if decorated by MnO nanoparticles for the fabrication of high-performing anodes for flexible LIBs.

5.2.2. Supercapacitors

Supercapacitors (SCs) are able to store impressively higher energy density than traditional capacitors, thanks to the creation of an electric double layer (EDL) as the key mechanism to store the charges. Moreover, SCs are able to accept and deliver charges quite faster than batteries and with a quite higher durability to charge/discharge cycles than batteries. Due to these features, SCs are usually considered as the technology filling the gap between conventional capacitors and LIBs [51]. The use of carbon-based nanofibers to design electrodes for supercapacitors is strictly related to high interest to develop flexible, portable and easy-to-integrate SCs for smart electronics. As already introduced discussing flexible LIBs, carbon-based nanofibers combine unique mix of electrical conductivity, mechanical flexibility and electrochemical stability that make a great material to design flexible systems. These impressive properties are coupled to the high versatility offered by the electrospinning technique that makes relatively easy to prepare composite and decorated carbon nanofibers [52].

5.3. Fuels

Hydrogen is progressively gaining importance as one of the possible green fuels of the future, able to become a potential substitute for fossil fuels. The electrocatalysis of the hydrogen evolution reaction (HER) is the critical step for this technology, pushing intense research to identify high-performing and low-cost catalysts. The availability of catalyst, able to efficiently drive the HER, while keeping the overall costs of the process, is indeed the mandatory requirement for large-scale H_2 production by this technology. In the area of electrochemical water splitting, carbon-based nanofibers by electrospinning are frequently used as a conductive and robust matrix to offer a support for other catalysts. Recently the huge potential of composite carbon nanofibers has been considered in this area too. Zhao et al. synthetized N-doped carbon nanofibers with embedded Co nanoparticles. They demonstrated the superior electrocatalytic performance of the resulting electrode, explaining the results as due to the ability of the material to expose two catalytic sites, the nitrogen defects into the main carbon nanofibers and the Co nanoparticles [53]. Again with the aim to identify possible substitute to expensive noble metal-based catalysts, Chen et al. optimized an interesting method to directly synthetized WO3-x in carbon-based nanofibers. They improve the thermal treatment to induce graphitization of the starting electrospun nanofibers, as well as the synthesis of the oxygen vacancy-rich WO3-x [54]. In recent years a new interest emerged in developing efficient electrochemical processes for the conversion of the environmental harmful CO_2 into new high-value products. Efficient catalysts are needed to grant good conversion efficiency, possibly involving low-cost materials that can help making competitive the process. In this area carbon-based nanofibers obtained by the pyrolysis of PAN nanofibers have been successfully investigate as catalysts of CO_2 reduction into CO. Kumar et al. demonstrated that nitrogen defects play a crucial role in catalyzing the reaction, with a mechanism quite similar to the one demonstrated for the catalysis of the ORR by the same class of materials [55].

6. Summary

The ever-increasing energy demand, related to the progress of human activities, favored an intense scientific research for the development of new technologies able to harness, convert and store environmental energy with safer and more efficient approaches than the traditional ones. In this frame, electrochemical processes for energy conversion have emerged as a unique, versatile and reliable platform to design efficient energy systems. This chapter has shown the key advancements obtained in the area of EEC by the introduction of the electrospinning process to design a new generation of carbon-based electrodes starting from nonwoven nanofiber mats. Relevant examples of electrodes made of carbon-based nanofibers have been demonstrated in all the key EEC technologies, from photovoltaics to batteries, from fuel cells to supercapacitors, clearly showing the most promising strategies introduced up to now.

Author details

Giulia Massaglia[1,2*] and Marzia Quaglio[1*]

*Address all correspondence to: giulia.massaglia@polito.it and marzia.quaglio@iit.it

1 Center For Sustainable Future Technologies@Polito, Istituto Italiano Di Tecnologia, Torino, Italy

2 Department of Applied Science and Technology (DISAT) Politecnico di Torino, Torino, Italy

References

[1] Kirubakaran A, Jain S, Nema RK. A review on fuel cell technologies and power electronic interface. Renewable and Sustainable Energy Reviews. 2009;**13**:2430-2440

[2] Bianco S, Chiodoni A, Nair JR, Gerbaldi C, Quaglio M. Nanostructures for energy. In: Bhushan B, editor. Encyclopedia of Nanotechnology. Dordrecht: Springer; 2016. pp. 2813-2827

[3] González A, Goikolea BJA, Mysyk R. Review on supercapacitors: Technologies and materials. Renewable and Sustainable Energy Reviews. 2016;**58**:1189-1206

[4] Gigot A, Fontana A, Serrapede M, Castellino M, Bianco S, Armandi M, Bonelli B, Pirri CF, Tresso E, Rivolo P. Mixed 1T–2H phase MoS2 /reduced graphene oxide as active electrode for enhanced Supercapacitive performance. ACS Applied Materials and Interfaces. 2016;**8**:32842-32852

[5] Krivik P, Baca P. Electrochemical energy storage. In: Zobaa AF, editor. Energy Storage – Technologies and Applications. London: IntechOpen; 2013

[6] Roger I, Shipman MA, Symes MD. Earth-abundant catalysts for electrochemical and photoelectrochemical water splitting. Nature Reviews Chemistry. 2017;**1**. DOI: 10.1038/s41570-016-0003

[7] Piumetti M, Fino D, Russo N. Photocatalytic reduction of CO2 into fuels: A short review. Journal of Advanced Catalysis Science and Technology. 2014;**1**:16-25

[8] Inagaki M, Yang Y, Kang F. Carbon nanofibers prepared via electrospinning. Advanced Materials. 2012;**24**:2547-2566. DOI: 10.1002/adma.201104940

[9] Huang ZM, Zhang YZ, Kotaki M, Ramakrishna S. A review on polymer nanofibers by electrospinning and their applications in nanocomposites. Composites Science and Technology. 2003;**63**:2223-2253. DOI: 10.1016/S0266-3538(03)00178-7

[10] Teo WE, Ramakrishna S. A review on electrospinning design and nanofibre assemblies. Nanotechnology. 2006;**17**:89-106. DOI: 10.1088/0957-4484/17/14/R01

[11] Bhardwa N, Kundu SC. Electrospinning: A fascinating fiber fabrication technique. Biotechnology Advances. 2010;**28**:325-347. DOI: 10.1016/j.biotechadv.2010.01.004

[12] Feng C, Khulbe KC, Matsuura T. Recent progress in the preparation, characterization, and applications of nanofibers and nanofiber membranes via electrospinning/interfacial polymerization. Journal of Applied Polymer Science. 2010;**115**:756-776. DOI: 10.1002/app.31059

[13] Hohman M, Ruletdge G, Brenner M. Electrospinning and electrically forced jets. II. Applications. Physics of Fluids. 2001;**13**:2221-2236. DOI: 10.1063/1.1384013

[14] Shin Y, Hohmann M, Brenner M, Ruletdge G. Experimental characterization of electrospinning: The electrically forced jet and instabilities. Polymer. 2001;**42**:9955-9967. DOI: 10.1016/S0032-3861(01)00540-7

[15] Reneker D, Yarin A, Fong H, Koombhongse S. Bending instability of electrically charged liquid jets of polymer solutions in electrospinning. Journal of Applied Physics. 2000;**87**:4531-4547. DOI: 10.1063/1.373532

[16] Deitzel J, Kleinmeyer J, Harris D, Tan NCB. The effect of processing variables on the morphology of electrospun nanofibers and textiles. Polymer. 2001;**42**:261-272. DOI: 10.1016/S0032-3861(00)00250-0

[17] Bella F, Massaglia G, Chiodoni A, Pirri FC, Quaglio M. Dispelling clichés at the nanoscale: The true effect of polymer electrolytes on the performance of dye-sensitized solar cells. Nanoscale. 2015;**7**:12010-12017. DOI: 10.1039/C5NR02286J

[18] Megleski S, Stephens JS, Chase DB, Rabolt JF. Micro-and nanostructured surface morphology on electrospun polymer fibers. Macromolecules. 2002;**35**:8456-8466. DOI: 10.1021/ma020444a

[19] Rahaman MSA, Ismail AF, Mustafa A. A review of heat treatment on polyacrylonitrile fiber. Polymer Degradation and Stability. 2007;**92**:1421-1432. DOI: 10.1016/j.polymdegradstab.2007.03.023

[20] Manickam SS, Karra U, Huang L, Bui NN, Li B, McCutcheon JR. Activated carbon nanofiber anodes for microbial fuel cells. Carbon. 2013;**53**:19-28. DOI: 10.1016/j.carbon.2012.10.009

[21] Yordem OS, Papila M, Menceloglu YZ. Effect of electrospinning parameters on polyacrylonitrile nanofiber diameter: An investigation by response surface methodology. Materials & Design. 2008;**29**:34-44. DOI: 10.1016/j.matdes.2006.12.013

[22] Liu J, Wang PH, Li RY. Continuous carbonization of polyacrylonitrile-based oxidized fibers: Aspects on mechanical properties and morphological structure. Journal of Applied Polymer Science. 1994;**52**:945-950. DOI: 10.1002/app.1994.070520712

[23] Bashir Z. A critical review of the stabilisation of polyacrylonitrile. Carbon. 1991;**29**:1081-1090. DOI: 10.1016/0008-6223(91)90024-D

[24] Dalton S, Heatley F, Budd PM. Thermal stabilization of polyacrylonitrile fibres. Polymer. 1999;**40**:5531-5543. DOI: 10.1016/S0032-3861(98)00778-2

[25] Ko TH. Influence of continuous stabilization on the physical properties and microstructure of PAN-based carbon fibers. Journal of Applied Polymer Science. 1991;**42**:1949-1957. DOI: 10.1002/app.1991.070420719

[26] Deng L, Young RJ, Kinloch IA, Zhu Y, Eichhorn SJ. Carbon nanofibres produced from electrospun cellulose nanofibers. Carbon. 2013;**58**:66-75. DOI: 10.1016/j.carbon.2013.02.032

[27] Chen D, Liu T, Zhou X, Tjiu WC, Hou H. Electrospinning fabrication of high strength and toughness polyimide nanofiber membranes containing multiwalled carbon nanotubes. The Journal of Physical Chemistry. B. 2009;**113**:9741-9748. DOI: 10.1021/jp9025128

[28] Lv YY, Wan LS, Xu ZK. Novel Porphyrinated polyimide nanofibers by electrospinning. Journal of Physical Chemistry C. 2008;**112**:1069-10615. DOI: 10.1021/jp7105549

[29] Kim WJ, Chang JY. Molecularly imprinted polyimide nanofibers prepared by electrospinning. Materials Letters. 2011;**65**:1388-1391. DOI: 10.1016/j.matlet.2011.02.010

[30] Kim C, Choi YO, Lee WJ, Yang KS. Supercapacitor performances of activated carbon fiber webs prepared by electrospinning of PMDA-ODA poly(amic acid) solutions. Electrochimica Acta. 2004;**50**:883-887. DOI: 10.1016/j.electacta.2004.02.072

[31] Yang KS, Edie DD, Lim DY, Kim YM, Choi YO. Preparation of carbon fiber web from electrostatic spinning of PMDA-ODA poly(amic acid) solution. Carbon. 2003;**41**:2039-2046. DOI: 10.1016 /S0008-6223(03)00174-X

[32] Xuyen NT, Ra EJ, Geng HZ, Kim KK, An KH, Lee YH. Enhancement of conductivity by diameter control of polyimide-based electrospun carbon nanofibers. Journal of Physical Chemistry B Letters. 2007;**111**:11350-11353. DOI: 10.1021/jp075541q

[33] Kim JR, Choi SW, Jo SM, Lee WS, Kim BC. Electrospun PVdF-based fibrous polymer electrolytes for lithium ion polymer batteries. Electrochimica Acta. 2004;**50**:69-75. DOI: 10.1016/j.electacta.2004.07.014

[34] Yang Y, Centrone A, Chen L, Simeon F, Hatton A, Rutledge GC. Highly porous electrospun polyvinylidene fluoride (PVDF)-based carbon fiber. Carbon. 2011;**49**:3395-3403. DOI: 10.1016/j.carbon.2011.04.015

[35] Zhou L, Gan L, Kang F, Wang M, Shen W, Huang Z. Sn/C non-woven film prepared by electrospinning as anode materials for lithium ion batteries. Journal of Power Sources. 2010;**195**:1216-1220. DOI: 10.1016/j.jpowsour.2009.08.052

[36] Son WK, Youk JH, Lee TS, Park WH. Effect of pH on electrospinning of poly (vinyl alcohol). Materials Letters. 2005;**59**:1571-1575. DOI: 10.1016/j.matlet.2005.01.025

[37] Zussman E, Yarin AL, Bazilevsky V, Avrahami R, Feldman M. Electrospun polyacrylonitrile/poly(methyl methacrylate)-derived turbostratic carbon micro-/nanotubes. Advanced Materials. 2006;**18**:348-353. DOI: 10.1002/adma.200501153

[38] Kim C, Jeong YI, Ngoc BTN, Yang KS, Kojima M, Kim YA, Endo M, Lee JW. Synthesis and Characterization of porous carbon nanofibers with hollow cores through the thermal treatment of electrospun copolymeric nanofiber webs. Small. 2007;**3**:91-95. DOI: 10.1002/smll.200600243

[39] Pant D, Singh A, Van Bogaert G, Olsen SI, Nigam PS, Diels L, Vanbroekhoven K. Bioelectrochemical systems (BES) for sustainable energy production and product recovery from organic wastes and industrial wastewaters. RSC Advances. 2012;**2**:1248-1263

[40] Logan BE, Rabaey K. Conversion of wastes into bioelectricity and chemicals by using microbial electrochemical technologies. Science. 2012;**337**:686

[41] Chen S, Hou H, Harnisch F, Patil SA, Carmona-Martinez AA, Agarwal S, Zhang Y, Sinha-Ray S, Yarin SA, Greiner A, Schroder U. Electrospun and solution blown three-dimensional carbon fiber nonwovens for application as electrodes in microbial fuel cells. Energy and Environmental Science. 2011;**4**:1417

[42] Nagasawa K, Takao S, Higashi K, Nagamatsu S, Samjeske G, Imaizumi Y, Sekizawa O, Yamamoto T, Uruga T, Iwasawa Y. Performance and durability of Pt/C cathode catalysts with different kinds of carbons for polymer electrolyte fuel cells characterized by electrochemical and in situ XAFS techniques. Physical Chemistry Chemical Physics. 2014;**16**:10075-10087

[43] Muthuswamy N, Buan MEM, Walmsley JC, Rønning JC. Evaluation of ORR active sites in nitrogen-doped carbon nanofibers by KOH post treatment. Catalysis Today. 2018;**301**:11-16

[44] Qiao Q. Carbon nanostructures as low cost counter electrode for dye-sensitized solar cells. In: Kosyachenko LA, editor. Solar Cells – Dye-Sensitized Devices. London: IntechOpen; 2011

[45] Joshi P, Zhang L, Chen Q, Galipeau D, Fong H, Qiao Q. Electrospun carbon nanofibers as low-cost counter electrode for dye-sensitized solar cells. ACS Applied Materials and Interfaces. 2010;**2**:3572-3577

[46] Hieu NT, Baik SJ, Jun Y, Lee M, Chung OH, Park JS. Electrospun coaxial titanium dioxide/carbon nanofibers. Electrochimica Acta. 2014;**142**:144-151

[47] Bianco S, Chiodoni A, Nair JR, Gerbaldi C, Quaglio M. Nanostructures for energy. In: Bhushan B, editor. Encyclopedia of Nanotechnology. Dordrecht: Springer; 2016. pp. 2813-2827

[48] Ji L, Toprakci O, Alcoutlabi M, Yao Y, Li Y, Zhang S, Guo B, Lin Z, Zhang X. α-Fe2O3 nanoparticle-loaded carbon nanofibers as stable and high-capacity anodes for rechargeable Lithium-ion batteries. ACS Applied Materials and Interfaces. 2012;**4**:2672-2679. DOI: 10.1021/am300333s

[49] Zhu J, Lei D, Zhang D, Li G, Lu B, Wang T. Carbon and graphene double protection strategy to improve the SnOx electrode performance anodes for lithium-ion batteries. Nanoscale. 2013;**5**:5499-5505. DOI: 10.1039/c3nr00467h

[50] Samuel E, Jo HS, Joshi B, An S, Park HG, Kim YI, Yoon WY, Yoon SS. Decoration of MnO nanocrystals on flexible freestanding carbon nanofibers for Lithium ion battery anodes. Electrochimica Acta. 2017;**231**:582-589. DOI: 10.1016/j.electacta.2017.02.077

[51] Wu Z, Li L, Yan J, Zhang X. Materials design and system construction for conventional and new-concept supercapacitors. Advanced Science. 2017;**4**:1600382

[52] Iqbal N, Wang X, Babar AA, Zainab G, Yu J, Ding B. Flexible Fe3O4@carbon nanofibers hierarchically assembled with MnO2 particles for high-performance supercapacitor electrodes. Scientific Reports. 2017;**7**:15153

[53] Zhao Y, Zhang J, Li K, Ao Z, Wang C, Liu H, Sun K, Wang G. Electrospun cobalt embedded porous nitrogen doped carbon nanofibers as an efficient catalyst for water splitting. Journal of Materials Chemistry A. 2016;**4**:12818

[54] Chen JD, Yu DN, Liao WS, Zheng MD, Xiao LF, Zhu H, Zhang M, Du ML, Yao JM. WO3−x nanoplates grown on carbon nanofibers for an efficient Electrocatalytic hydrogen evolution reaction. ACS Applied Materials and Interfaces. 2016;**8**:18132-18139

[55] Kumar B, Asadi M, Pisasale D, Sinha-Ray D, Rosen BA, Haasch R, Abiade J, Yarin AL, Salehi-Khojin A. Renewable and metal-free carbon nanofibre catalysts for carbon dioxide reduction. Nature Communications. 2013;**4**:2819. DOI: 10.1038/ncomms3819

Chapter 8

Electrospun Nanofibers for Entrapment of Biomolecules

Diána Balogh-Weiser, Csaba Németh, Ferenc Ender,
Benjámin Gyarmati, András Szilágyi and
László Poppe

Additional information is available at the end of the chapter

http://dx.doi.org/10.5772/intechopen.76068

Abstract

This chapter focuses on nanofiber fabrication by electrospinning techniques for the effective immobilization of biomolecules (such as enzymes or active pharmaceutical ingredients—APIs). In this chapter, the development of precursor materials (from commercial polymer systems to systematically designed biopolymers), entrapment protocols, and precursor-nanofiber characterization methods are represented. The entrapment ability of poly(vinyl alcohol) and systematically modified polyaspartamide nanofibers was investigated for immobilization of two different lipases (from *Candida antarctica* and *Pseudomonas fluorescens*) and for formulation of the antibacterial and antiviral agent, rifampicin. The encapsulated biomolecules in electrospun polymer fibers could be promising nanomaterials for industrial biocatalysis to produce chiral compound or in the development of smart drug delivery systems.

Keywords: electrospinning, entrapment, enzyme immobilization, drug delivery

1. Introduction

Recently, a considerable effort has been focused on nanofiber fabrication, nanofibrous materials, and the application of such materials. The characteristic features of nanofibers are such as small diameter (within the 100 nm–1 μm range), large specific surface area, infinite length, and high aspect ratio [1]. These properties make nanofibers suitable for a wide range of applications including, but not limited to, medical applications, cancer cell engineering, tissue engineering, drug delivery systems, enzyme immobilization, and electronics [2]. Ceramics, metals, and polymers are used to fabricate nanofibers. In fact, the fabrication techniques can limit the utilization of the nanofibers. Therefore, constant attention is paid to the improvement of the existing fabrication techniques and developing novel fabrication methods. Among various methods, electrospinning is the most widely used process for nanofiber fabrication [1]. Recently, polymer nanofibers have gained more and more attention in development of "bio-engineered" or "bio-inspired" systems for pharmaceutical, biomedical, or biotechnological applications. Among these widespread issues, tissue engineering for artificial tissue reconstruction and replacement, smart drug formulations for targeted drug delivery, bioselection processes for selective filtering and sensoring materials, as well as biocatalyst production by enzyme or whole-cell immobilization are the most highlighted areas (**Figure 1**). This chapter focuses on biocatalyst design and drug delivery systems by utilizing the entrapping ability of polymer nanofibers.

Polymer nanofibers can be used to immobilize both small and macromolecules by their physical adsorption or covalent binding on the surface of the fibers or by entrapment within the fiber. The choice between these two different possibilities depends on the application and on the type of small/macromolecules that would be immobilized. The main benefits and disadvantages of the attachment and entrapment are compared in **Table 1**. Generally, entrapment of biomolecules is the more beneficial way to immobilize biocatalysts (such as enzymes or whole cells) or encapsulate drugs or vitamins, due to the significant stabilizing and protective effect and the controllable retention and release of the entrapped molecules.

In the process of electrospinning, a polymer solution held by its surface tension at the end of a capillary tube is subjected to an electric field [3, 4]. For entrapment of a biomolecule (small or a macro-sized) by electrospinning, formation of a homogenous precursor mixture from the biomolecule and polymer solution is required, which can be continuously fed by a syringe pump in an electrostatic field with high voltage. Two electrodes are used: one electrode is at the end of a capillary fed by the precursor mixture and the other is attached to a collector. By applying high voltage on one electrode (usually in the range of 10 to 30 kV), while the collector electrode is grounded, strong electrostatic field develops, therefore charge is induced on the liquid surface. Mutual charge repulsion causes a force directly opposite to the force arising from the surface tension. Increasing intensity of electrostatic field elongates

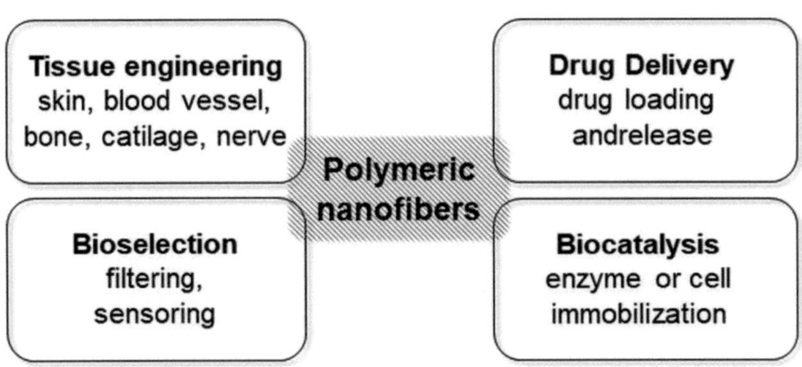

Figure 1. Application fields of nanofibers as functional biomaterials.

	Attachment onto polymer nanofiber	Entrapment within polymer fibers
Advantages	• The biomolecule and the precursor solution can be handled separately • No solvent or excipient materials limitation during electrospinning	• Easy to prepare • In situ and rapid immobilization/formulation • Significant stabilizing effects • Controllable liberation/retention of the biomolecule
Disadvantages	• (Difficult) chemical surface modification steps are necessary • Immobilization can be achieved step by step	• Chemical compatibility between the precursor system and biomolecule can be a limiting factor

Table 1. Comparison of the attachment and entrapment of biomolecules onto or within polymer nanofibers.

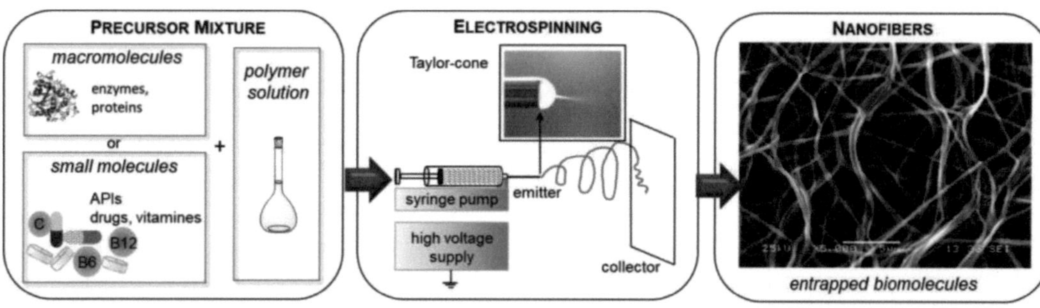

Figure 2. Entrapment of biomolecules by electrospinning technique.

the hemispherical surface of the solution at the tip of the capillary tube and forms a conical shape known as the Taylor cone. When the electric field reaches a critical value—at which the repulsive electric force overcomes the force from surface tension—a charged jet of the mixture is ejected from the tip of the Taylor cone [3]. Since this jet is charged, it will move toward the collector, and its trajectory can be controlled by an applied electric field. As the jet travels in air, the solvent evaporates, and the polymer fibers remain on a collector surface. Thus, completing the electrospinning process, continuous fibers are formed that result in a fibrous material comprising entrapped biomolecules (**Figure 2**). Experimental results presented in this chapter were carried out by the electrospinning equipment eSpin (Spinsplit Ltd., Budapest, Hungary).

2. Nanobiocatalysts: entrapment of enzymes in electrospun nanofibers

In the simplest sense, biocatalysis can be defined as the use of biological systems to catalyze (speed up) chemical reactions. These substances of natural origin can be one or more enzymes in isolated form or within whole cells. An enzyme being a protein catalyst which accelerates the reaction of target molecules under mild conditions can be used in various industries. Major limiting factors of the application of such biocatalysts in various industries are the sensitivity of enzymes to environmental conditions and the high cost of their manufacturing. This is why they ought to be protected from detrimental conditions until their use and—if possible—they should be recycled for subsequent usage in order to reduce the costs. The benefits of nanofibrous structures can be advantageously applied in these processes. Nanofibers can be used as a protective carrier for enzymes and the large surface area of nanofibers allows rapid release of the biomolecules when and where needed. Moreover, enzymes can also be immobilized on or within electrospun nanofibers for repeated use. Due to the unique properties of nanofibers, such as extraordinarily high surface area and tunable surface morphology, high permeability, low density, ability to retain electrostatic charges, and cost effectiveness, they can be ideally applied for enzyme immobilization. In our previous study, it was found that electrospun poly(vinyl alcohol) (PVA) and poly (lactic acid) (PLA) nanofibers were applicable for entrapment of lipase from *Burkholderia cepacia* (lipase PS), Pseudomonas fluorescens (lipase AK) and lipase B from Candida antarctica (CaLB). The lipase PS and CaLB biocatalysts entrapped in PVA nanofibers were durable biocatalysts retaining significant part of their original biocatalytic activity after ten cycles [5, 6].

2.1. Enzyme immobilization capacity of electrospun polymer nanofibers

One of the key issues during enzyme entrapment is to reach the optimal enzyme loading with the highest specific activity of the valuable enzyme. To optimize enzyme loading in the course of entrapment, nanofiber entrapment of lipase AK into PVA and of CaLB into a cationic polyaspartamide were examined at different enzyme/polymer ratios. The immobilized lipase biocatalysts were tested in kinetic resolution of racemic 1-phenylethanol (*rac*-**1**, **Figure 3**) using vinyl acetate as acylating agents in *n*-hexane/*t*-butyl methyl ether (MTBE). The biocatalytic properties of the different electrospun biocatalysts, such as specific enzymatic activity (U_E, U/g) and enantiomeric excess (*ee*, %), were determined from gas chromatographic analysis of the reaction media.

In case of lipase AK-PVA nanofibers, the specific enzyme activity (U_E) reached the highest level when fibers were loaded with 5 and 20% enzyme (**Figure 4a**). In contrast, with CaLB in polyaspartamide, the specific enzyme activity showed a clear tendency with a maximum at moderate enzyme loading. Similar to lipase AK in PVA, 5% enzyme content exhibited the maximal specific activity (**Figure 4b**).

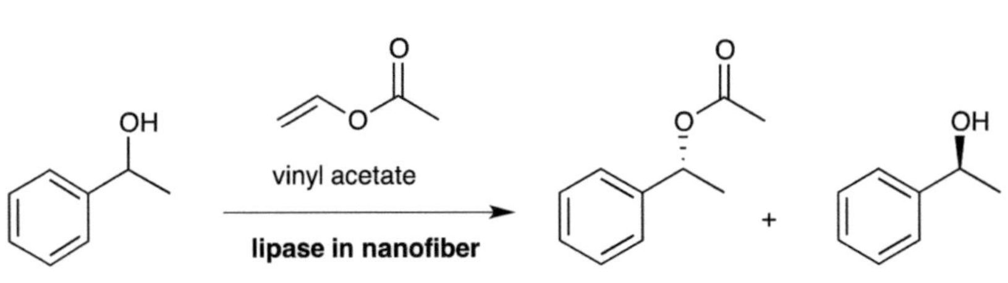

rac-**1** (*R*)-**2** (*S*)-**1**

Figure 3. Kinetic resolution of racemic 1-phenylethanol (*rac*-1) catalyzed by lipase biocatalysts entrapped within electrospun nanofibers.

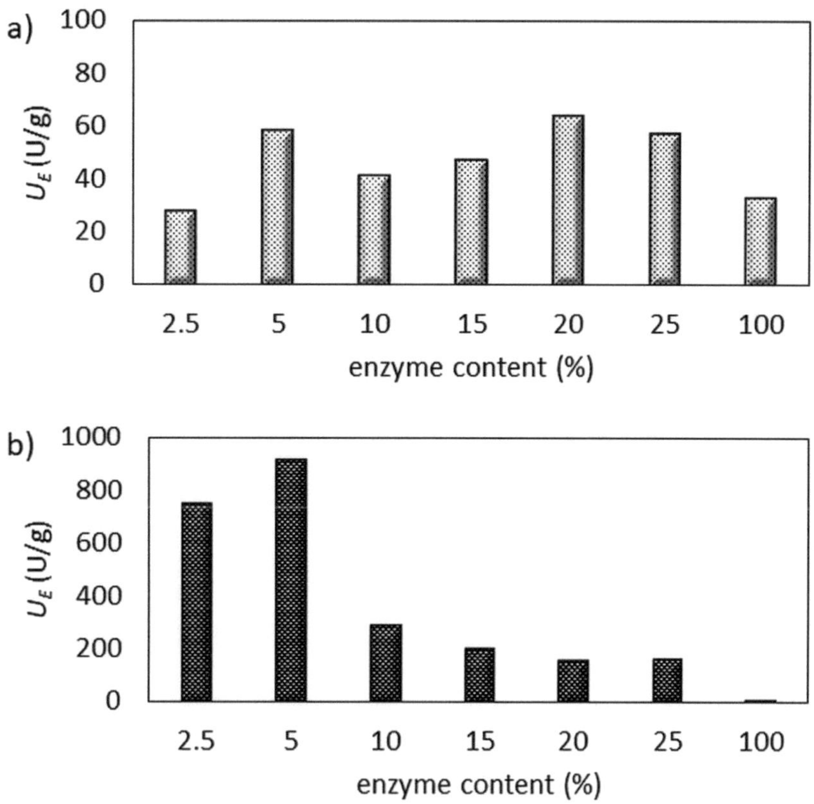

Figure 4. Kinetic resolution of 1-phenylethanol (*rac*-**1**) with (a) lipase AK entrapped in PVA nanofibers and (b) CaLB entrapped in polyaspartamide nanofibers at different enzyme loading.

The morphological properties of the nanofibers with different enzyme content were investigated by scanning electron microscopy (SEM). In case of lipase AK in PVA nanofibers, SEM images showed that at 2.5 and 5% enzyme content (**Figure 5a** and **b**) the nanofibers were uniform with consistently homogenous surfaces. However, at enzyme content higher than 5%, more and more inhomogeneity could be observed, and nanofibers were not continuous (**Figure 5c–f**).

Similarly to lipase AK-PVA systems, in case of CaLB-polyaspartamide nanofibers, the morphology of fibers depended strongly on the enzyme content. CaLB at higher than 5% loading resulted in significant heterogeneity of the nanofibers (**Figure 6c–f**).

In summary, enzyme activity tests as well as SEM images of the biocatalyst demonstrated clearly that finding the optimal enzyme content of the nanofibers is a key issue. In case of lipase AK entrapped in PVA and also for CaLB in polyaspartamide nanofibers, the 5% enzyme content was optimal to reach maximal specific activity of the entrapped enzyme and homogenous nanofiber structure. With both lipases and polymer matrices, enzyme contents higher than 5% resulted in formation of heterogeneous nanotissues with bead-like shapes and forming from aggregates of polymer or enzyme molecules. On the one hand, application of the optimal enzyme loading enables the most economical and cost-effective

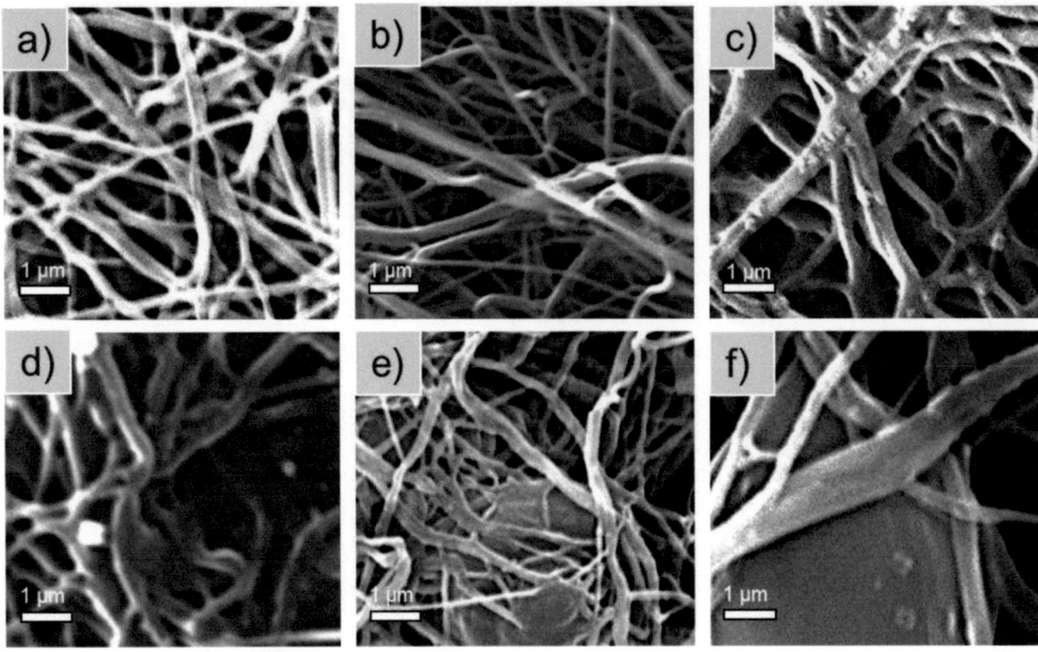

Figure 5. Scanning electron microscope images of PVA nanofibers loaded with different amounts of lipase AK: (a) 2.5, (b) 5, (c) 10, (d) 15, (e) 20, (f) 25%.

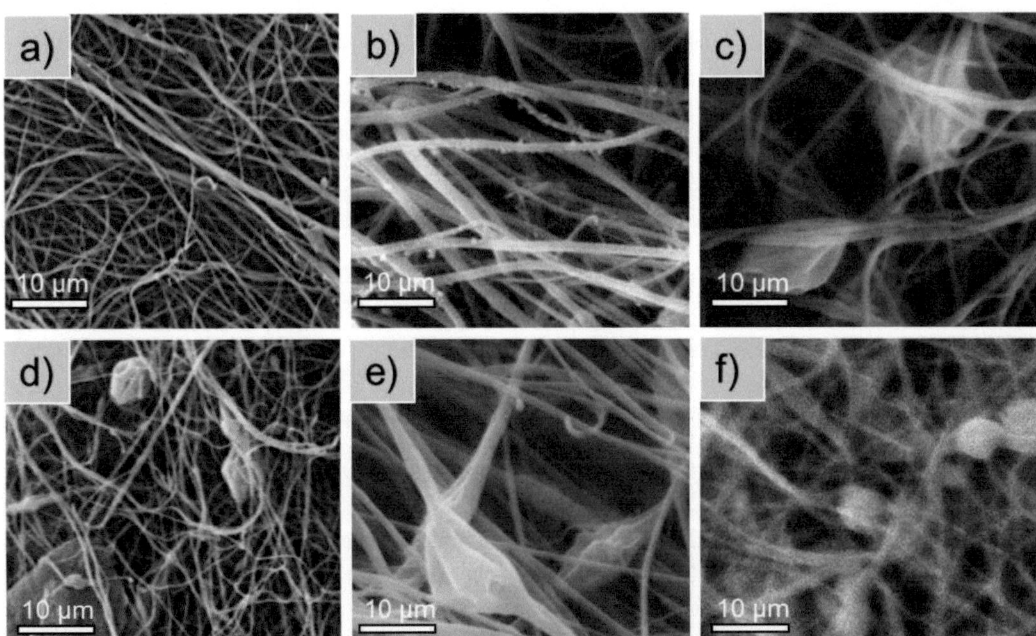

Figure 6. Scanning electron microscope images of polyaspartamide nanofibers loaded with different amounts of CaLB: (a) 2.5, (b) 5, (c) 10, (d) 15, (e) 20, (f) 25%.

usage, which is quite important at an industrial scale. In addition, homogenous fiber morphology is crucial to produce high-quality nanobiocatalysts in a reproducible manner.

2.2. Bioimprinting for enhanced biocatalytic activity of enzyme entrapped in nanofibers

One of the most promising possibilities for enhancing enzyme activity during immobilization, especially for entrapment methods, is molecular imprinting which conserves by the aid of added substrates or substrate analogues the proper shape of the enzyme's active site [7, 8]. This so-called bioimprinting, as can be rationalized by the generally accepted hypothesis of the interfacial activation mechanism, can influence the biocatalytic activity and enantioselectivity of lipases. The positive effect of bioimprinting on lipases was already demonstrated by immobilizations using sol-gel entrapment [9, 10]. The active site of many lipases in aqueous solution is covered by a flexible region of the enzyme, often referred to as a lid. Interaction of the lid with hydrophobic molecules can enforce its opening to make the active site accessible [11, 12]. This hypothesis is supported by crystal structures of lipases in their open and closed forms [13, 14]. Because interfacial activation can increase significantly the catalytic activity and selectivity of lipases, it should be considered for all applications of lipases including development of novel immobilization methods. Although bioimprinting proved to be an efficient tool to modulate the properties of lipases entrapped in sol-gel matrices, this strategy has been extended for entrapment in electrospun nanofibers only recently in one of our previous studies [7].

To demonstrate the efficiency and generality of bioimprinting during entrapment of lipases, lipase AK and CaLB were entrapped in PVA and polyaspartamide nanofibers in the presence of four different additives as bioimprinting agents. Poly(ethylene glycol)s (PEGs) of different molecular weights and nonionic detergents such as Tween 80 and Brij 30 were already applied as substrate analogues exhibiting bioimprinting effects in sol-gel systems [11] and in PVA nanofibers [6], but their effect in polyaspartamide entrapment has not been investigated yet. To reveal the effect of additives and the difference between fiber-forming polymer matrices, CaLB was chosen as model enzyme for the bioimprinting experiments. Results of testing the immobilized nanofibrous CaLB biocatalysts by kinetic resolution of *rac*-**1** showed that all tested additives increased the specific enzyme activity (U_E) and the enantiomeric excess of the product (ee_P) (**Table 2**). The

Additive	**CaLB-PVA**		**CaLB-polyaspartamide**	
	U_E	ee_P	U_E	ee_P
	(U/g)	**(%)**	**(U/g)**	**(%)**
–	208	99.8	1470	99.8
PEG 400	294	99.9	2621	99.9
PEG 1000	204	99.5	2690	99.7
Tween 80	228	99.9	2262	99.9
Brij 30	180	99.8	2563	99.9

Table 2. Kinetic resolution of 1-phenylethanol (rac-**1**) with CaLB-PVA or CaLB-polyaspartamide nanofibers fabricated in the presence of different additives.

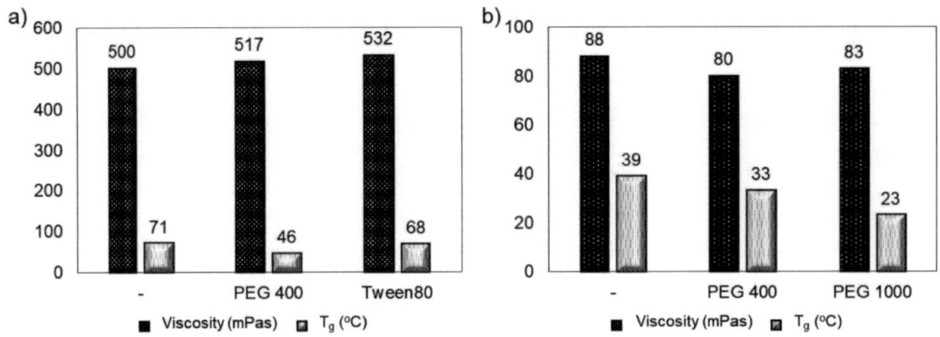

Figure 7. Dynamic viscosity and glass transition temperature (T_g) of electrospun nanofibers (a) from PVA- and (b) from polyaspartamide-based precursor systems.

best results were obtained with PEG 400 and Tween 80 as additives in electrospinning of CaLB-polyaspartamide fibers, while, in case of electrospinning of CaLB-polyaspartamide fibers, PEG 400 and PEG 1000 were the most efficient bioimprinting agents. It must be noted, that CaLB in polyaspartamide polymer matrix without additive exhibited much higher activity than in PVA. Consequently, additives could manifest their beneficial effect on the activity of CaLB in polyaspartamide nanofibers better than in PVA nanofibers.

The polymer matrix of nanofibers is able to influence the enzyme activity by the bioimprinting effect. In addition, the physicochemical properties of the matrix material may affect also significantly the apparent enzyme activity and the final properties of immobilized biocatalyst. Interactions between polymer chains can affect significantly the diffusion limitations for substrate or product, which can strongly influence the apparent efficiency of the immobilized biocatalyst. Thus, dynamic viscosity as a rheological property of the precursor systems (enzyme-polymer-additive mixtures) and glass transition temperature (T_g) as a thermal property of the resulted enzyme-filled nanofibrous materials were also investigated (**Figure 7**). According to Kramers' theory, the biocatalytic activity of enzymes could strongly depend on the viscosity of solvent because viscosity results in friction of proteins in solution leading to decreased motion and inhibiting catalysis by motile enzymes [15]. Thus, viscosity of the precursor system could affect directly the properties of the immobilized enzyme during its entrapment. Moreover, the viscosity of the polymer precursor system can significantly influence fiber formation during electrospinning as well. The dynamic viscosity values of PVA- and polyaspartamide-based precursor systems were determined in the presence of CaLB and the best additives such as PEG 400, PEG 1000, and Tween 80 and were compared to the simple polymer solution without the enzyme or additive. The glass transition temperature of the electrospun products was investigated as well. In case of CaLB-PVA systems, PEG 400 slightly increased the viscosity but Tween 80 resulted in higher increase in the viscosity value. On the other hand, when T_g values of the different CaLB-PVA fibrous biocatalysts were compared, PEG 400 had more pronounced effects on the interaction between the polymer chains, while the T_g of PEG 400-containing CaLB-PVA fibers was significantly lower than that of the CaLB-PVA fibers without additives (**Figure 7a**). In case of CaLB-polyaspartamide, additives (PEG 400 and PEG 1000) had no significant effect on the viscosity of precursor system, but T_g values became lower due to the presence of PEG molecules, especially in the case of PEG 1000 (**Figure 7b**).

3. Entrapment of drugs in electrospun nanofibers

3.1. Encapsulation of rifampicin in polyaspartamide nanofibers for ophthalmic applications

The interest in the development of novel ophthalmic drug formulations has increased considerably because of the low bioavailability of drug molecules after their administration on the surface of the eye. The low therapeutic efficiency is caused by the complex structure of the eye, the small absorptive surface, and low transparency of the cornea, lipophilicity of corneal epithelium, metabolism, bonding of the drug to proteins in tear liquid, and protective mechanisms such as tear formation, blinking, and the flow of the active pharmaceutical ingredients (APIs) through the nasolacrimal duct [16, 17]. The main challenge in the development of ophthalmic drug formulations is to achieve the required drug concentration at the site of absorption and to improve residence time, which in turn contributes to smaller application frequency [18, 19].

Polymers are often used in ophthalmic formulations to increase viscosity and thus, residence time of the formulation on the corneal surface, which increases the bioavailability of the API. Drug penetration can also be enhanced by various additives such as chelating agents, surfactants, and cyclodextrins, which form inclusion complexes leading to increased solubility, permeability, and bioavailability of the poorly soluble drugs. Controlled release of the drug can also be achieved by novel formulations such as inserts, collagen shields, contact lenses, and in situ gels. These solid or semi-solid formulations provide controlled release and increasing bioavailability of the API due to the extended contact time with cornea. Electrospun nanofibers offer controllable release rate as well as exact and safe dosage of the API.

In our work, a novel solid nanofibrous formulation was designed to encapsulate rifampicin (RFP), an important antibiotic and antiviral agent. Rifampicin inhibits bacterial DNA-dependent RNA synthesis; thus, it can be applied for the treatment of many infectious diseases such as tuberculosis and Hansen's disease [20]. Although rifampicin has great efficiency, the scope of its applications is limited by its poor hydrolytic and thermal stability and by its rather limited solubility [21]. Due to these difficulties and the inaccurate dosing of eye drops, development of a stable solid formulation of rifampicin is still required.

Poly(aspartic acid) (PASP) is a biocompatible and biodegradable synthetic poly(amino acid) having great potential in biomedical applications. For using as a polymer in electrospinning, PASP derivatives, including cationic polyaspartamides, can be synthesized under mild reaction conditions [22, 23]. The biological activity of the PASP-based polymers has been investigated carefully and, owing to their protein-like structure, they are expected to be biodegradable. Due to these advantageous properties of polyaspartamide derivatives, rifampicin can be beneficially entrapped in polyaspartamide-based nanofibers fabricated by electrospinning. Poly(aspartic acid) derivatives with butyl and hexyl groups were tested to determine the effect of chemical structure on release properties. For electrospinning of PASP derivatives containing rifampicin (1.6 w/w%), ethanol was used as the solvent to avoid the degradation of the active compound. The morphology of electrospun fibers was investigated by SEM indicating that uniform nanofibers could be produced independently of chemical composition. By using butyl side groups, the fiber diameter was ~800 nm, while

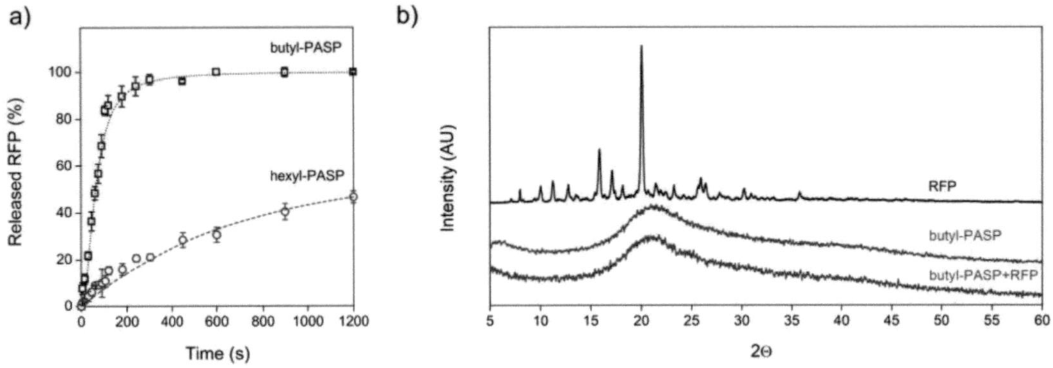

Figure 8. (a) Release of rifampicin (RFP) from butyl-PASP and hexyl-PASP electrospun matrices and (b) X-ray diffraction spectra of free rifampicin, butyl-PASP electrospun matrix, and rifampicin entrapped in butyl-PASP.

in the presence of hexyl groups, the average diameter of the fibers was ~900 nm. The results of release experiments (**Figure 8a**) showed that the dissolution rate of rifampicin was larger by using butyl-PASP than hexyl-PASP, but in both cases, the total amount of the encapsulated drug was released.

The crystallinity of entrapped API determines its bioavailability; thus, X-ray diffractograms were recorded for the matrices and the native drug. From the comparison of the X-ray spectra of native and encapsulated rifampicin (butyl-PASP + RFP), it can be clearly seen that the originally crystalline rifampicin became amorphous during electrospinning, which can improve the bioavailability even further (**Figure 8b**).

4. Conclusion

Electrospinning is a promising tool for efficient entrapment of biomolecules, such as expensive enzymes or active pharmaceutical ingredients, within polymer nanofibers which are applicable for biotechnology or biomedical purposes. By rational selection of polymer excipients, both small and macromolecules can be immobilized into electrospun matrices and controllable retention and release can be achieved. Poly(vinyl alcohol) (PVA) and polyaspartamide nanofibers are promising candidates to encapsulate enzymes. In case of industrially relevant lipases, such as lipase from *Candida antarctica* (CaLB) or *Pseudomonas fluorescens* (lipase AK), the optimal enzyme loading of nanofibers generated by electrospinning was 5% entrapped in PVA or polyaspartamide matrices. Entrapment of lipases in nanofibers in the presence of different substrate-like additives led to bioimprinting, which could significantly increase the activity and enantioselectivity of the entrapped enzyme. Polyaspartamides could also be used to encapsulate active pharmaceutical ingredients by electrospinning. The encapsulation of rifampicin proved that the chemical modification of PASP with different alkyl side groups had a significant effect on the release rate of APIs, opening up new possibilities for the improvement of therapeutic efficiency with solid dosage forms.

Acknowledgements

This research was supported by the National Research, Development, and Innovation Office—NKFIH (FK 125074). Benjámin Gyarmati acknowledges the János Bolyai Research Scholarship of the Hungarian Academy of Sciences for financial support. A. Szilágyi is grateful to the support of the ÚNKP-17-4-III New National Excellence Program of the Ministry of Human Capacities. We thank Anna Szabó for her cooperation in the electrospinning of polyaspartamides (Department of Physical Chemistry and Materials Science, Budapest University of Technology and Economics, BME) and Imre Miklós Szilágyi (Department of Inorganic and Analytical Chemistry, BME) for performing the XRD measurements.

Conflict of interest

The authors declare no conflict of interest.

Author details

Diána Balogh-Weiser[1,2,3], Csaba Németh[2], Ferenc Ender[4,5], Benjámin Gyarmati[2], András Szilágyi[2] and László Poppe[1,3,6]*

*Address all correspondence to: poppe@mail.bme.hu

1 Department of Organic Chemistry and Technology, Budapest University of Technology and Economics, Budapest, Hungary

2 Department of Physical Chemistry and Materials Science, Budapest University of Technology and Economics, Budapest, Hungary

3 SynBiocat Ltd., Budapest, Hungary

4 Department of Electron Devices, Budapest University of Technology and Economics, Budapest, Hungary

5 Spinsplit Ltd., Budapest, Hungary

6 Faculty of Chemistry and Chemical Engineering, Biocatalysis and Biotransformation Research Centre, Babeş-Bolyai University of Cluj-Napoca, Cluj-Napoca, Romania

References

[1] Xue J, Xie J, Liu W, Xia Y. Electrospun nanofibers: New concepts, materials, and applications. Accounts in Chemical Research. 2017;**50**:1976-1987. DOI: 10.1021/acs.accounts.7b00218

[2] Doshi J, Reneker DH. Electrospinning process and applications of electrospun fibers. Journal of Electrostatics. 1995;**35**:151-160. DOI: 10.1016/0304-3886(95)00041-8

[3] Yarin AL. Taylor cone and jetting from liquid droplets in electrospinning of nanofibers. Journal of Applied Physics. 2001;**90**:4836. DOI: 10.1063/1.1408260

[4] Ramakrishna S, Fujihara K, Teo WE, Lim TC, Ma Z. Electrospinning process. In: An Introduction to Electrospinning and Nanofibers. Singapore: World Scientific Publishing Co.; 2005. pp. 90-154. DOI: 10.1142/9789812567611_0003

[5] Sóti P, Weiser D, Vígh T, Nagy Z, Poppe L, Marosi G. Electrospun polylactic acid and polyvinyl alcohol fibers as efficient and stable nanomaterials for immobilization of lipases. Bioprocess and Biosystems Engineering. 2016;**39**:449-459. DOI: 10.1007/s00449-015-1528-y

[6] Weiser D, Sóti P, Bánóczi G, Bódai V, Kiss B, Gellért Á, Nagy ZK, Koczka B, Szilágyi A, Marosi G, Poppe L. Bioimprinted lipases in PVA nanofibers as efficient immobilized biocatalysts. Tetrahedron. 2016;**72**:7335-7342. DOI: 10.1016/j.tet.2016.06.027

[7] Dickey FH. The preparation of specific adsorbent. Proceedings of the National Academy of Sciences of the United States of America. 1949;**35**:227-229. http://www.pnas.org/content/35/5/227

[8] Mingarro I, Abad C, Braco L. Interfacial activation-based molecular bioimprinting of lipolytic enzymes. Proceedings of the National Academy of Sciences of the United States of America. 1995;**92**:3308-3312. DOI: 10.1073/pnas.92.8.3308

[9] Cao X, Yang J, Shu L, Yu B, Yan Y. Improving esterification activity of *Burkholderia cepacia* lipase encapsulated in silica by bioimprinting with substrate analogues. Process Biochemistry. 2009;**44**:177-182. DOI: 10.1016/j.procbio. 2008.10.003

[10] Hellner G, Boros Z, Tomin A, Poppe L. Novel sol-gel lipases by designed bioimprinting for continuous-flow kinetic resolutions. Advanced Synthesis & Catalysis. 2011;**353**: 2481-2491. DOI: 10.1002/adsc.201100329

[11] Verger R. Interfacial activation' of lipases: Facts and artifacts. Trends in Biotechnology. 1997;**15**:32-38. DOI: 10.1016/S0167-7799 (96)10064-0

[12] Reis P, Holmberg K, Watzke H, Leser ME, Miller R. Lipases at interfaces: A review. Advances in Colloid Interface. Sciences. 2009;**147-148**:237-250. DOI: 10.1016/j.cis.2008.06.001

[13] Brady L, Brzozowski AM, Derewenda ZS, Dodson E, Dodson G, Tolley S, Turkenburg JP, Christiansen L, Hugejensen B, Norskov L, Thim L, Menge U. Nature. 1990;**343**:767-770. DOI: 10.1038/343767a0

[14] Schrag JD, Li Y, Wu S, Cygler M. Ser-His-Glu triad forms the catalytic site of the lipase from Geotrichum candidum. Nature. 1991;**351**:761-764. DOI: 10.1038/351761a0

[15] Uribe S, Sampedro JG. Measuring solution viscosity and its effect on enzyme activity. Biological Procedures. 2003;**5**:108-115. DOI: 10.1251/bpo52

[16] O'Connor Davies H, Hopkins GA, Pearson RM. The Actions and Uses of Ophthalmic Drugs. 3rd ed. London: Butterworths; 1989. ISBN: 9781483192048

[17] Pahuja P, Arora S, Pawar P. Ocular drug delivery system: A reference to natural polymers. Expert Opinion on Drug Delivery. 2012;**9**:837-861. DOI: 10.1517/17425247.2012.690733

[18] Karthikeyan D, Bhowmick M, Pandeyetal VP. The concept of ocular inserts as drug delivery systems: An overview. Asian Journal of Pharmaceutics. 2008;**2**:192-200. DOI: 10.22377/ajp.v2i4.204

[19] Kumari A, Sharma PK, Garg VK, G. Ocular inserts—Advancement in therapy of eye diseases. Journal of Advanced Pharmaceutical Technology and Research. 2010;**1**:291-296. DOI: 10.4103/0110-5558.72419

[20] Calvori C, Frontali L, Leoni L, Tecce G. Effect of rifamycin on protein synthesis. Nature. 1965;**207**:417-418. DOI: 10.1038/207417a0

[21] Shishoo CJ, Shah SA, Rathod IS, Savale SS, Kotecha JS, Shah PB. Stability of rifampicin in dissolution medium in presence of isoniazid. International Journal of Pharmaceutics. 1999;**190**:109-123. DOI: 10.1016/S0378-5173(99)00286-0

[22] Krisch E, Gyarmati B, Szilágyi A. Preparation of pH-responsive poly(aspartic acid) nanogels in inverse emulsion. Periodica Polytechnica-Chemical Engineering. 2017;**61**:19-26. DOI: 10.3311/PPch.9788

[23] Németh C, Szabó D, Gyarmati B, Gerasimov AV, Varfolomeev M, László K, Szilágyi A. Effect of side groups on the properties of cationic polyaspartamides. European Polymer Journal. 2017;**93**:805-814. DOI: 10.1016/j.eurpolymj.2017.02.024